# GENIUS

# INVENTIONS

Published in 2019 by André Deutsch
An imprint of the Carlton
Publishing Group
20 Mortimer Street
London WIT 3JW

This edition is revised and adapted from
*Genius: Great Inventors and their Creations* (2010).

Text © Jack Challoner 2010, 2019
Design © André Deutsch 2019

A CIP catalogue record for this book is available from the British Library.

ISBN  978 0 233 00539 3

Printed in Dubai

10 9 8 7 6 5 4 3 2 1

Cover illustrations: Science & Society Picture Library & Shutterstock

# GENIUS INVENTIONS

## THE STORIES BEHIND HISTORY'S GREATEST TECHNOLOGICAL BREAKTHROUGHS

JACK CHALLONER

ANDRE
DEUTSCH

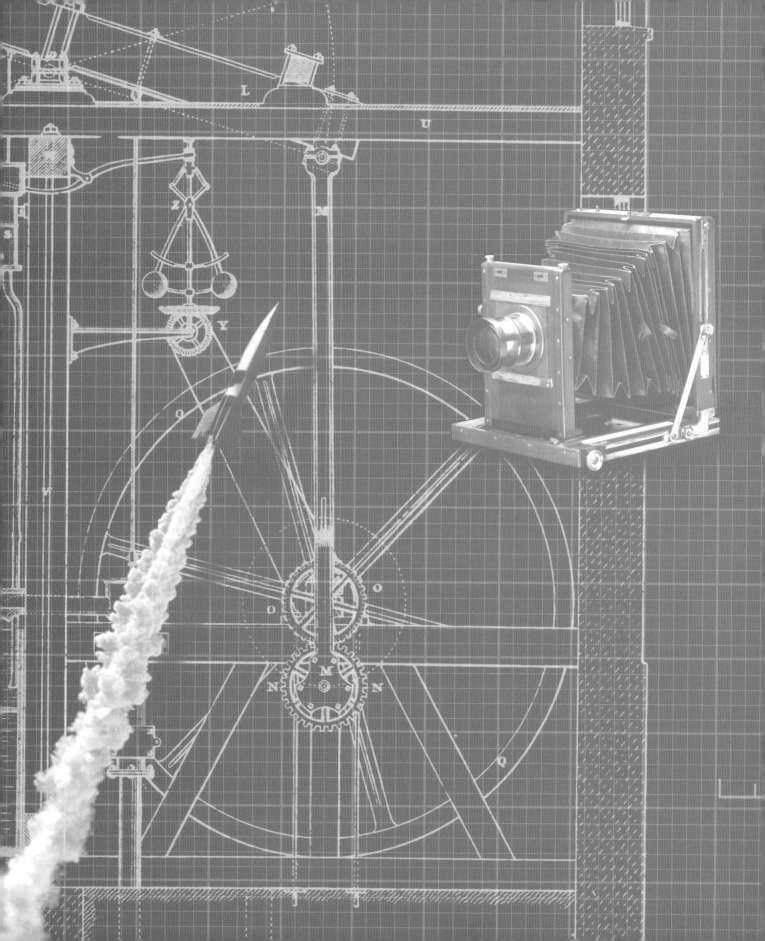

# Contents

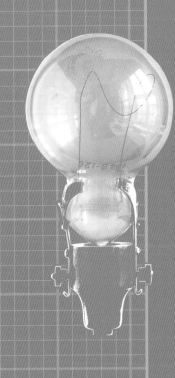

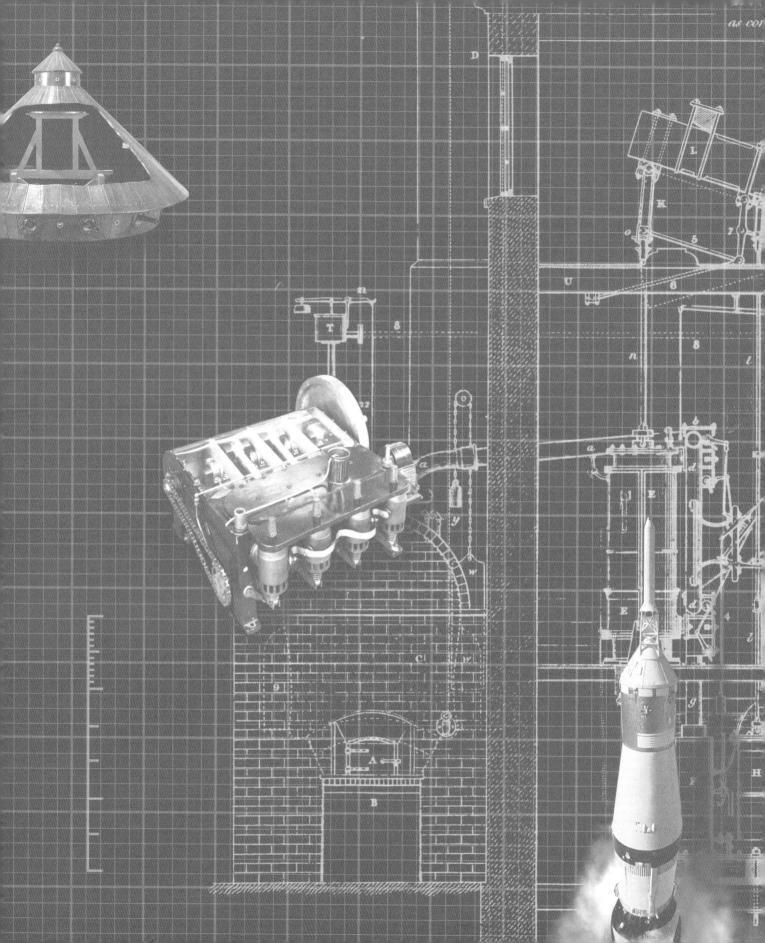

# Introduction

THIS BOOK IS A CELEBRATION OF THE BRILLIANT DISCOVERIES AND
INNOVATIONS THAT HELPED SHAPE THE MODERN WORLD.
THEY ARE BY NO MEANS THE ONLY GREAT DEVELOPMENTS IN HISTORY, NOR ARE THE
INVENTORS MENTIONED THE ONLY NOTEWORTHY ONES, BUT THE 28 SUBJECTS IN
THIS BOOK ARE CRUCIAL TO THE HISTORY OF SCIENCE AND TECHNOLOGY.

*Genius Inventions* is organized chronologically by the dates of
the first forays into each discovery, beginning in *c.*287 BCE.
Of course, the history of invention does not begin here, with
Archimedes. You can trace its beginnings back to the first stone
tools fashioned by our distant ancestors in Africa more than two
million years ago. More recently, Old Stone Age (Palaeolithic)
people invented weaponry, fire-making and clothing; and
around eleven thousand years ago, New Stone Age (Neolithic)
people in the Fertile Crescent, in Mesopotamia and the Levant,
began farming and building houses. The great civilizations that
followed – in ancient Mesopotamia, China, India and Egypt –
introduced many fundamental technologies. Innovations such
as the wheel, bricks, boats, ploughs and the smelting of metals
are still of great importance today, but in most cases, little or
nothing is known of the background.

Next, we note the great contribution by early Islamic
scholars to the history of science and technology, before moving
on to Renaissance Europe, where the seeds of the modern
world were sown. In the 1440s, Johannes Gutenberg brought
together a number of existing technologies to invent the printing
press, which quickly spread new ideas and encouraged literacy.
Renaissance polymath Leonardo da Vinci's technological genius
was perhaps too far ahead of its time, and as a result, his work
had less influence on the development of technology than it
could have had. Scientific instruments such as the telescope,
the microscope and the thermometer, invented in the sixteenth
century, were crucial in the "Scientific Revolution" that gave
people a desire, and the means, to understand of the world.
As science matured, it began to play an increasingly important
role in the process of invention.

Science was also important in the Industrial Revolution,
which began in Britain in the 1750s. It helped usher in a
"mechanistic" view of the world, which helped advance

engineering and contributed to the development of the
steam engine. High-pressure steam went on to power railway
locomotives, bringing about a revolution in transport.
Discoveries in physics, chemistry and biology underpinned
most of the other important advances of the nineteenth
century, including the electric motor and generator,
photography, and antiseptic surgery. Many of the inventions
that characterize the modern world were developed between
1870 and 1930, including the telephone, the motor car,
electrification, cinema, the aeroplane, and more. Nearly half
the inventors featured in this book were active in that period,
which historians often refer to as the "Second Industrial
Revolution". But innovation has not faltered since then: the
second half of the twentieth century saw rockets reach the
Moon, the rise of electronic computers, tremendous advances
in medicine and the invention of the World Wide Web.

The story of human ingenuity, which began some 6,000
years ago, is never ending. Scanning the following pages, it
quickly becomes clear how many of the things that, today,
most people take for granted were the result of out-of-the-box
thinking by a fascinating variety of people from all walks
of life. And, though most early inventors were men – their
advances occurred in rich, industrialized countries, in societies
which those people privileged with education and opportunity
were mostly white men – what is also true is that, as time
elapsed, women began to have their many efforts recognized.
Windscreen wipers, the first computer language and key
components of the Mars Rovers were all devised by women.

Today, things are very different and more and more genius
inventors have the access and the platform to build their
ideas – although there is still a long way to go before everyone
everywhere has similar opportunities to create technology that
can change the world.

# Ancient Inventions

## PEOPLE HAVE BEEN INVENTING THINGS SINCE THE DAYS OF CAVEMEN. USUALLY, WE DON'T KNOW WHO CAME UP WITH VERY EARLY CONCEPTS BECAUSE NO ONE THOUGHT TO WRITE ABOUT THEM.

One of the problems with knowing who was responsible for ancient inventions lies with the ability to record such events. Ancient China had seen experiments with writing materials for a couple of hundred years BCE, but it was a court eunuch of the imperial Han Dynasty who, in 75 CE, first began macerating tree bark, hemp, rags and fishnets to create felted sheets of paper. Better (and cheaper) than the silk currently in use, Cai Lun's papermaking process was seized upon by Emperor He and the idea was swiftly adopted throughout first China, then the rest of the world. Silk itself had far earlier origins. According to the Chinese philosopher Confucius, it was discovered by Empress Hsi Ling-Shi in roughly 2640 BCE. As well as finding out how to cultivate silk worms, she is thought to have been responsible for the invention of the silk reel and loom.

We are aware of Archimedes's (c.287 BCE–c.212 BCE) mathematical achievements from his own writings, but any knowledge of his remarkable inventions exists only because his contemporaries documented them. Archimedes was born in Syracuse, on the island of Sicily, then a colony of the Grecian Empire. The little that is known about his life comes from commentaries written by historians who lived at the time or over the next hundred or so years. The most important source is Greek-born Greek and Roman biographer and historian Plutarch (c.46–120 CE).

According to Plutarch, Archimedes's father was an astronomer and the family was closely related to the ruler of Syracuse, King Hiero II (c.306–215 BCE). King Hiero asked Archimedes to design a pump to drain his ship during the voyage to Alexandria, Egypt. Archimedes devised a simple, yet brilliant, solution. Today known as the Archimedes Screw (or Archimedean Screw), his device consists of a helical blade – a wide screw thread – inside a cylinder. The screw lifts water when it turns and was so effective that it was quickly adopted

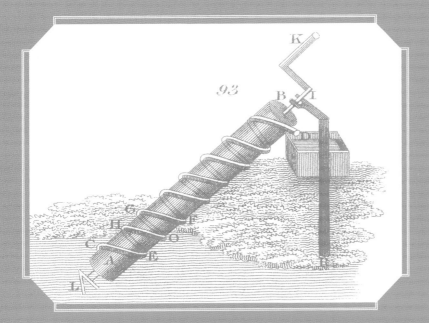

**LEFT:** An 1815 print showing the inside of an Archimedes Screw, normally housed in a cylinder. Turning the handle clockwise drags the water up the screw thread, through the cylinder, so that it emerges at the top. The device was used extensively for irrigation in Archimedes's day, and brought him great fame.

**OPPOSITE:** An ancient Roman mosaic uncovered early in the nineteenth century during French excavations of Pompeii, Italy. It shows Archimedes at his table with an abacus.

in many countries for irrigation. Archimedean Screws are still used worldwide for the same purpose and are commonplace in factories and on earth-moving machines, where they are used to move granular materials such as soil and plastic pellets.

Archimedes studied the mathematics of the day in Alexandria but quickly moved beyond it, realising the close and important connection between mathematics and experimental and mechanical principles. His exquisite mathematical proofs and inspired ideas that reveal his true genius.

Although none of Archimedes's original work in his own hand exists, several copies made during the first thousand or so years after his death still exist. The most important is an eleventh-century manuscript on vellum. Archimedes's work had been scraped off, overwritten with Christian prayers, and bound together as part of a book. The book, now called *The Archimedes Palimpsest*, was bought at auction in 1998 and scientists have been applying the latest imaging techniques to "see through" the Christian text. One of the most remarkable findings is that Archimedes invented some of the principles of the mathematical technique today called calculus. Crucial to modern science and technology, calculus was only actually formalized in the late seventeenth century, by Isaac Newton (1643–1727) and Gottfried Leibniz (1646–1716).

9

**RIGHT:** *The Archimedes Palimpsest* – a book of Christian prayers (horizontal) written in the twelfth century over a tenth-century copy of some of Archimedes's most important works (vertical). Scientists at the Walters Art Museum in Baltimore, USA, have used a variety of techniques to make the Archimedes text more visible.

## BLOCK AND TACKLE

Ancient civilizations made use of what physicists call "simple machines": the lever, the ramp, the wheel and axle, the inclined plane, the wedge and the pulley. Archimedes was almost certainly the first to combine two pulleys to make a device that could exert a huge force.

According to Plutarch, Archimedes invented the block and tackle after suggesting to King Hiero that there is no weight too great to be moved by a lever. Hiero challenged Archimedes to move the heavy ship *Syracusia*, a feat normally only achieved by teams of many strong men. Archimedes single-handedly moved the ship, complete with crew and cargo, with his new invention. That device, the block and tackle, is still used today for lifting or pulling heavy loads.

Archimedes used his knowledge of gears to invent a small, wheeled cart that could measure long distances (an odometer), a clock that struck the hours, and devices to predict the positions of the Sun, the Moon and the then-known five planets.

In 1900, divers discovered what scholars deduced was an ancient astronomical computer in a wreck off the coast of the Greek island Antikythera. Some historians believe this computer may be closely descended from the work of Archimedes.

Of all Archimedes's inventions, the ones most celebrated in his lifetime were the weapons he designed to defend Syracuse during the siege of the city by the Romans, which began in 214 BCE. This included the Claw – a crane fixed to the city wall that could lift Roman ships out of the water and drop or capsize them. Plutarch wrote three slightly different accounts of Archimedes's last moments, but all state that Archimedes died during the Siege of Syracuse, killed by the sword of a Roman soldier.

Archimedes was not a lone mathematician, but it could be argued that Hypatia (c.355 CE–415 CE) was unique, at least in her gender. The earliest well-documented female mathematician, she was the daughter of celebrated mathematician and astronomer Theon. Hypatia continued her father's work, eventually becoming head of the Platonist school at Alexandria. Studying and teaching philosophy, she was also well-versed in practical matters and the latest inventions. Letters survive asking her advice on the construction of an astrolabe and hydroscope. Early extremist Christians, however, felt threatened by her level of scholarship and she was murdered in 415 CE by an Alexandrian mob.

Around 800 years earlier, a chemist called Tapputi (c.1,200 BCE) was creating perfume using the world's first-recorded chemical process. An ancient Babylonian tablet, written in cuneiform, tells us Tapputi-Belatekallim ('female palace overseer') and her (female) assistant distilled the essence of flowers and other fragrant materials, filtering and returning them to the still as many times as it took to perfect her recipe. We do not know that Tapputi actually invented the still, but it is the first reference to the concept. Tapputi wrote a treatise about the process and although the text is now lost, we still have one of her recipes, a healing balm that includes myrrh, oil, flowers and the root of the calamus plant.

**OPPOSITE:** Part of the Antikythera mechanism, which appears to be an ancient astronomical calculator and was recovered from the wreck of a Roman ship dating back to the first century BCE. Archimedes is known to have built devices for this purpose and many academics believe this could be one of them.

## TIMELINE

**c.2,600,000 BCE**
Human ancestors, *Homo habilis*, first make stone tools.

**c.10,000 BCE**
Domestication of animals and the first agriculture, in the Fertile Crescent, in and around modern-day Turkey. The move from hunter-gatherer to settled lifestyle encouraged experimentation and the spread of new technologies.

**c.6500 BCE**
Smelting of metals; the earliest-known extraction of metals from their ores is evidenced by lead beads found in the early settlement of Çatalhöyük, in modern-day Turkey. Gold and copper naturally exist in pure form, and were used to make tools around 7000 BCE.

**c.4000 BCE**
The plough evolves from simple hand-held hoes to larger devices dragged by animals, resulting in better crop yields for less effort. The wheel is invented in Mesopotamia, first as a potters' wheel, and then used on carts.

**c.3000 BCE**
Glass is invented – almost certainly by accident – somewhere in Mesopotamia or Egypt. The earliest man-made objects were transparent glass beads, which have been found in both these areas.

**c.1500 BCE**
Bronze shears invented in Ancient Egypt were the first scissors.

**c.700 BCE**
The earliest known example of an arch bridge is built in Sumer, although arches were used in doorways as early as 1500 BCE.

**c.400 BCE**
The catapult, invented in Sicily, was not the first weapon – but it was the first mechanized one.

**c.300 BCE**
The water mill, invented in Ancient Greece, is the first example of people harnessing power other than human or animal muscle.

11

# The Islamic World

**MOST PEOPLE ARE AWARE OF THE TREMENDOUS SCIENTIFIC AND TECHNOLOGICAL ADVANCES OF THE GREAT ANCIENT CIVILIZATIONS IN EGYPT, CHINA, INDIA, GREECE AND ROME.**

But during the Middle Ages, the Islamic Empire kept the spirit of learning and innovation alive. One of its greatest technical geniuses was a mechanical engineer named Al-Jazarī (1136–1206). Badi' al-Zaman Abu al-'Izz Isma'il ibn al-Razzaz al-Jazarī was born in an area of Mesopotamia called al-Jazira, in what is now part of modern-day southern Turkey. Al-Jazarī lived at the height of the Islamic Golden Age, also sometimes called the Islamic Renaissance. The spread of Islam in the seventh century had encouraged a rich culture and a stable political system – the Caliphate. By 750 CE, the Caliphate covered a huge area, from northern Spain in the west, through the Middle East and North Africa, to the fringes of China in the east. Throughout this Islamic Empire, there was a great emphasis on learning; scholars collected and translated all the knowledge they could from around the world and added their own. From the ninth to the twelfth century, the Caliphate was the foremost intellectual centre of the world.

Out of the stability and the learning came great wealth, and powerful dynasties ruled over each region. Al-Jazarī became chief engineer to the Artuqid dynasty in the town of Diyarbakir, after his father retired from the same position in 1174. Most of what we know about al-Jazarī comes from a book he completed shortly before his death. The *Kitáb fi ma'rifat al-hiyal al-handasiyya* (*Book of Knowledge of Ingenious Mechanical Devices*) is a compendium of the engineering designs he created through his career. According to the book's introduction, Nasir al-Din Mahmud ibn Muhammad, the dynasty's ruler between 1200 and 1222, commissioned al-Jazarī to write the book in 1198.

**OPPOSITE:** Part of a page from *The Book of Knowledge of Ingenious Mechanical Devices*, where al-Jazarī depicted a container constructed to dispense a combination of four different wines.

**BELOW / LEFT:** Model of pump (initial design to the left), built for a 1976 exhibition called "Science and Technology in Islam" at the Science Museum, London, part of the countrywide Festival of Islam. The mechanisms are mostly hidden; most prominent is the waterwheel that would have driven the device.

Al-Jazarī's book contains details of 50 ingenious devices, including intricate clocks, fountains that regularly change their flow patterns, machines for raising water and toys for entertainment. The description of each device is accompanied by clear drawings that help explain how it was constructed and how it worked.

The spread of Islam brought huge advances in science, mathematics, medicine and philosophy. Engineering, on the other hand – although held in great esteem and practised competently – was mostly just a continuation of existing technologies established by the Greeks and the Romans. There were certainly notable exceptions, and some of those innovations are to be found in al-Jazarī's wonderful book. For example, al-Jazarī's water- or donkey-powered devices made use of power-transmission elements that had been used for centuries: gears, levers and pulleys. But in one of his inventions, a double-acting piston pump, he gives the first known reference to a crankshaft – a device for changing

**LEFT:** Reconstruction of Al-Jazarī's elephant clock at the Ibn Battuta Mall in Dubai, United Arab Emirates. Every half-hour, the scribe on the elephant's back rotates full circle, and at the end of each half-hour, the figure of the mahout (elephant driver) beats a drum and a cymbal sounds.

**BELOW:** Glass alembic, approximately eleventh century. An alembic is an essential tool in distillation, a procedure for purifying mixtures. Distillation was pioneered by Islamic chemists, who developed many processes that would later be important in the development of the science of chemistry.

rotary motion to back-and-forth motion (or vice versa). He also makes extensive use of the camshaft, a rotating cylinder with pegs protruding from it; his is the first mention of that, too. Al-Jazarī also invented the first known combination lock and the earliest known mechanical water-supply system, which was installed in Damascus in the thirteenth century, to supply hospitals and mosques across the city.

Several of al-Jazarī's contraptions featured automata: animal or human figures that made precise, programmed movements. For example, he describes a boat containing four automated musicians that entertained at parties and an automated girl figure that refilled a wash basin. Automatons also feature in most of al-Jazarī's clocks, which were more elaborate and ingenious than any that had come before. Most impressive was his "castle clock". More than 3 metres (10 feet) high, it displayed the constellations of the zodiac, with the orbits of the Sun and the Moon, and doors that opened every hour to reveal papier-mâché figures. This extraordinary device could also be programmed to take account of the varying day lengths.

**ABOVE:** Model of a blood-letting device described in al-Jazari's *Book of Knowledge of Ingenious Mechanical Devices*. Blood-letting (phlebotomy) was a popular practice in medieval Islamic medicine. This device measured the volume of blood lost during blood-letting sessions.

## TIMELINE

**700s** Muslim astronomers discover the astrolabe from ancient Greek texts and improve upon it. This calculator was the most important astronomical device until the invention of the telescope.

**750s** Abu Ja'far Abdallah ibn Muhammad al-Mansur (714–775) founds the Bayt al-Hikmah (House of Wisdom) in Baghdad, to collect and translate ancient Greek works.

**780s** Jābir ibn Hayyān, known in Europe as Geber (d.803), systematizes practice of alchemy, pioneering chemical techniques such as distillation and laboratory equipment including the alembic.

**820** Muhammad ibn Mūsā al-Khwārizmī (c.780–850) writes *Hisab al-jabr w'al-muqabala* (*Book on Calculation by Completion and Balancing*), the first book about algebra, which is translated into Latin in the 12th century.

**830s** At the House of Wisdom, Abū Ja'far al-Māʾmūn (786–833) sponsors translation of *Almagest* by Greek-Roman astronomer Ptolemy (c.90–c.168), a key text in Europe until the 16th century.

**1021** Ibn al-Haytham, better known as Alhazen (965–1040), completes his *Book of Optics*, highly influential in Europe.

**1025** Avicenna completes his *Canon of Medicine*, which becomes the standard medical textbook in most universities until the 16th century.

**1085** Christian forces seize the Muslim city of Toledo (in Spain), whose great library contains hundreds of Arabic translations of classic Greek texts.

**1160** Ibn Rushd, better known as Averroes (1126–1198), comments on and adds to the works of Aristotle, making them very popular in European universities.

**1175** Gerard of Cremona (c.1114–1187), from Lombardy (in Italy), translates Ptolemy's *Almagest* from Arabic to Latin, one of 87 translations he completed.

**1202** Italian mathematician Leonardo of Pisa, or Fibonacci (1170–1250), writes *Liber Abaci*, in which he presents to Europe the Hindu-Arabic numeral system.

15

16

## THE INFLUENCE OF ISLAMIC SCHOLARS

During the Islamic Golden Age, the centre of scholarly activity was the House of Wisdom in Baghdad (in modern Iraq). Both a library and a centre for translation, the House of Wisdom acted not only as a repository for the books and ideas of ancient thinkers from Greece and China, but also as a centre of excellence for contemporary scholars.

Much of the knowledge collected, translated and expanded by medieval Islamic scholars passed into Europe in the twelfth and thirteenth centuries. A dedicated band of European scholars sought out works in Spain and Sicily after these areas came under Christian rule. They translated what they found into Latin, and the resulting documents formed the basis of early scientific study in Europe.

The works of the Islamic scientists, mathematicians, astronomers and doctors contained significant advances in fields such as atomic theory, optics, surgery, chemistry and mathematics. Kept alive in the universities of medieval Europe, their ideas inspired the Scientific Revolution of the sixteenth and seventeenth centuries.

**ABOVE:** Persian scholar Abu Ali Ibn Sina, better known by his Latinized name Avicenna (*c.*980–1037 AD). Avicenna was a key figure in the transmission of classical Greek and Roman ideas to medieval Europe, but also contributed many of his own ideas and experiences, in more than 200 books.

**OPPOSITE:** Pages from *Book of Knowledge of Mechanical Devices*, completed in 1206, showing al-Jazari's ingenious devices. The book was illustrated – and several copies made – by members of a school of painting sponsored by the rulers of the Artuqid dynasty.

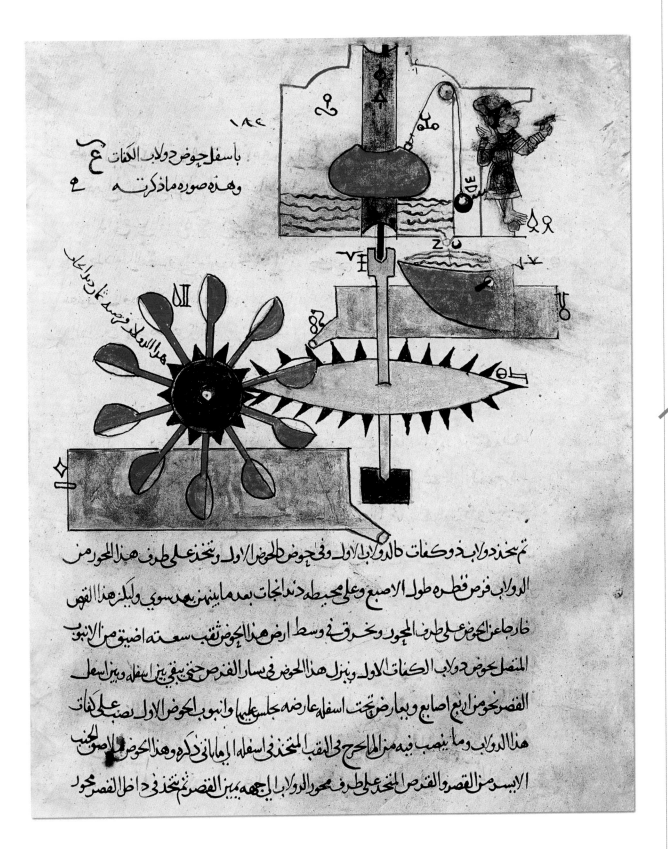

# Printing

## IT IS DIFFICULT TO OVERESTIMATE THE IMPORTANCE OF THE PRINTING PRESS IN THE HISTORY OF THE WORLD.

The mass-production of books made them cheaper and more accessible, which promoted literacy and the spread of ideas. The creator of this influential new technology was a German goldsmith named Johannes Gutenberg (c.1400–1468).

Little is known of the early life of Johannes (or Johann) Gutenberg. It is known that he was born in Mainz, Germany, around 1400, and that he came from the privileged, governing elite. He attended university, where he would have come into contact with books, and he trained as a goldsmith.

Around 1420, several families were exiled from Mainz after a rebellion by the tax-paying middle class. Gutenberg's was among them, and he travelled to Strasbourg, where he was involved in several ventures. One of them, he told his financial backers, was "a secret". It is very likely that this secret was the development of the printing press.

At the time, nearly all books were painstakingly written out by scribes. Books, therefore, were rare and extremely expensive, and literacy was confined to religious and political leaders. Woodblock printing produced a few books – but each block, representing a

**ABOVE:** A type case filled with large, decorative moveable type in a reconstruction of Gutenberg's printing workshop at the Gutenberg Museum in Mainz, Germany. A printer would slot these individual pieces of type into a frame, to represent the text of one page of a book.

**OPPOSITE:** Portrait of Johannes Gutenberg, 1584. Gutenberg's printing press enabled the rapid spread of new ideas. His most important invention was the hand mould, in which he cast copies of individual letters from an alloy of lead, tin and antimony.

whole page, had to be carved in its entirety. Gutenberg's important innovation, "moveable type", changed all that.

Moveable type is a system of printing in which a page of text is arranged in a frame, or matrix, by slotting in individual raised letters. The letters are then inked and pressed onto paper. It was invented in Korea and in China in the eleventh century, but never caught on, mostly because of the large number of characters that are used in written Chinese and Korean.

Gutenberg invented moveable type independently, and his approach was simple and efficient. First, he carefully made punches of hardened steel, each with the raised shape of a letter. With these, he punched impressions of the letters into copper. Next, he fitted the "negative" copper pieces into a hand-held mould of his own invention, and poured in molten metal to cast as many perfect copies of the letters as he needed. The metal Gutenberg used was an alloy of lead, tin and antimony that has a low melting point and solidified quickly inside the mould. His alloy is still used wherever "founder's type" or "hot metal" letterpress printing methods survive today.

While still in Strasbourg in the 1440s, Gutenberg experimented with another crucial element of his printing

**ABOVE:** Coloured nineteenth-century artist's impression of a scene in Gutenberg's workshop (artist unknown). Gutenberg is shown in the foreground, checking a printed page. There would actually have been about 20 people working in the workshop at any one time.

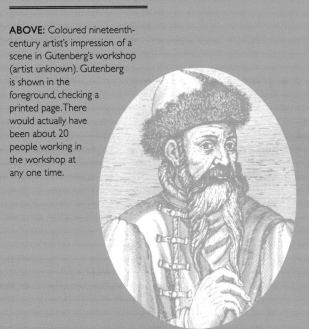

## ROTARY PRINTING PRESS

Although Gutenberg's invention dramatically changed the course of history in a very short time, printing was still a painstaking process. It required several people and produced only a hundred or so sheets per hour. The invention of cast-iron presses and the introduction of steam power in the nineteenth century improved that rate to about a thousand pages an hour. A further major step in the history of printing was the invention of the rotary press in 1843, by American inventor Richard March Hoe (1812–1886).

Hoe's steam-powered invention could print millions of pages per day, largely due to the fact that paper could be fed in through rollers as a continuous sheet. Hoe's device relied upon lithography, a process invented by Bavarian author Aloys Senefelder (1771–1834). In lithography, ink is applied to smooth surfaces rather than to raised type, which was ideally suited to the drum of Hoe's press.

**OPPOSITE TOP:**
The frontispiece of the oldest dated printed book. Bought from a monk in a cave in Dunhuang, China in 1907, this copy of the Buddhist text *Diamond Sutra* is on a scroll 5 metres (16 feet) long. It was printed using woodblocks in 868 CE.

**OPPOSITE BELOW:**
A highly decorated page from a Gutenberg Bible. Gutenberg produced 180 copies of his bible. Some were on vellum, others on paper; some were decorated (by hand), others were left plain. The books caused a sensation when they were first displayed at a trade fair in Frankfurt in 1454.

20

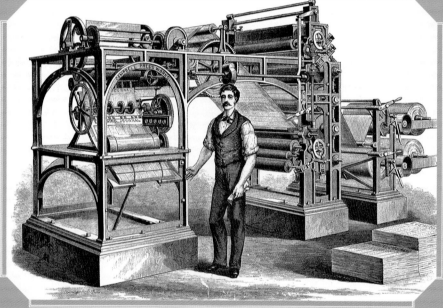

set up a printing shop there. Knowing that the church would be the main source of business, Gutenberg decided to print bibles. Work on the Gutenberg Bible began around 1452, after several test prints of other works, including books on Latin grammar. The relatively low price of the bibles, and their exquisite quality, secured the success of Gutenberg's new technology, which then spread quickly across Europe. By 1500, millions of books had been printed. Gutenberg had created the first media revolution.

system: the press. Gutenberg's press was adapted from winemakers' screw presses. The inked, typeset text was slotted face-up on a flat bed, covered with paper, then slid underneath a heavy stone; turning the screw then pressed the paper onto the type. Repeating the process gave exact copies time after time. Gutenberg also formulated oil-based ink, which was more durable than the water-based inks in use at the time. He knew that by putting all these technologies together he was onto something very important.

By 1448, Gutenberg was back in Mainz. He borrowed money from a wealthy investor, Johann Fust (c.1400–1466), to

Unfortunately for Gutenberg, Johann Fust demanded his money back, and accused Gutenberg of embezzlement. A judge ordered Gutenberg to hand over his printing equipment as payment. Fust went on to become a successful printer, and Gutenberg set up a smaller printing shop in the nearby city of Bamberg. Gutenberg later moved to a small village where, in 1465, he was finally recognized for his invention and given an annual pension. He died three years later in relative poverty.

## TIMELINE

**c.200 BCE** — Woodblock printing is invented in China. At the time, most printing is done on silk, but the invention of paper around the same time, also in China, makes the process easier.

**11th Century** — Chinese inventor Bi Sheng produces the first moveable type, using wooden letters. The first metal moveable type was used in Korea in the 13th century.

**1440s** — Johannes Gutenberg invents his printing press and independently invents moveable type, along with a way of making raised letters from a hand-held mould.

**1796** — German playwright Aloys Senefelder works out a way to print illustrations from a smooth, flat plate – a process called lithography. Initially, the plate was made of limestone, but later, metal plates were used.

**1818** — Senefelder invents chromolithography – an extension of lithography that enabled the printing of coloured illustrations.

**1843** — American inventor Richard March Hoe makes the first rotary press, which speeds up printing manifold.

**1855** — French chemist Alphonse Poitevin develops photolithography – a way of reproducing photographs in books and newspapers.

**1903** — American inventor Ira Rubel invents modern offset lithographic printing, in which a printed page is transferred, or "offset", onto a rubber roller before being printed onto paper.

**1969** — American researcher Gary Starkweather invents the laser printer, in which a laser casts an image onto a drum. The drum becomes electrically charged where the laser strikes, attracting toner ink.

**2000s** — Electronic paper becomes popular in e-book readers. Invented in the 1970s, electronic paper works with reflected light, like ordinary paper.

21

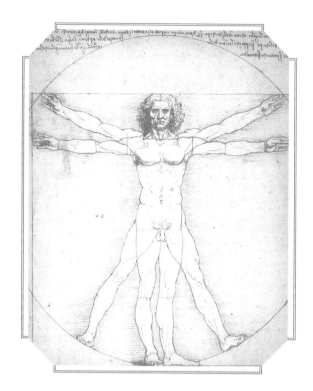

# Renaissance Science

## WHILE HIS OWN WORK WAS OFTEN THEORETICAL, THE WORK OF LEONARDO DA VINCI HAS HAD A PROFOUND INFLUENCE ON INVENTORS EVER SINCE.

Although undoubtedly a genius, the scientific research of Leonardo da Vinci (1452–1519), the archetypal Renaissance man, was not well known in his lifetime. Most of his inventions were never built and he was much more famous for his influence on painting, drawing and sculpture. A pioneer of perspective, he used anatomical studies to improve life drawing, found new ways to paint light and shade and used new materials and composition techniques. That Leonardo was also a great scientist, engineer and inventor only became common knowledge when his journals were published long after his death.

Born in Vinci, a town in Tuscany, Italy, Leonardo was apprenticed at sixteen, under the artist Andrea del Verrocchio (c.1435–1488) in Florence. He qualified as a master at the age of 20, and worked in Florence, then in Milan, where he created such iconic paintings as *The Adoration of the Magi, The Virgin of the Rocks* and *The Last Supper*.

Throughout his life, and particularly during his time in Milan, Leonardo kept detailed notebooks. There were an estimated 13,000 pages in all, containing his observations,

**ABOVE:** One of da Vinci's most famous drawings, *L'Uomo Vitruviano,* or *Vitruvian Man, c.*1490. It is based on the book *De architectura* by Roman architect Vitruvius (*c.*80–15 BC), who believed that the proportions of the human body were instrumental for creating proportions in the classic orders of architecture.

**OPPOSITE TOP, left:** Self portrait *c.*1510. Leonardo's supreme draughtmanship was in part due to his hands-on experience of anatomy.

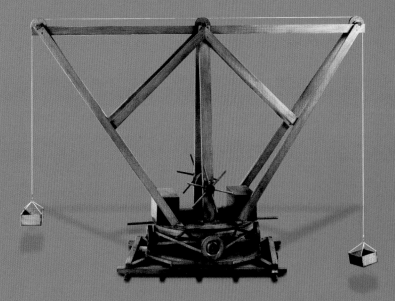

thoughts, sketches and inventions. Around 5,000 of these pages survive today. They reveal how Leonardo followed the scientific method – based on careful observation, scepticism and experiment – well before the likes of Galileo Galilei (1564–1642) and Isaac Newton (1643–1727). Leonardo's grasp of optics, geology, hydrodynamics (the behaviour of water), astronomy and the principles behind gears, levers, cantilevers and force and motion was far ahead of his time.

Leonardo had a chance to apply some of his knowledge and understanding while working as an engineer and military architect for two dukes of Milan from 1485 until 1499, and afterwards in the same capacity for other patrons, including the infamous Cesare Borgia (1475–1507). He made a point of promising his clients wonderful engineering projects, and only mentioned in passing that he was also a painter.

Interest in Leonardo's scientific work began to grow in the nineteenth century. His notebooks detailed plans for many incredible inventions, most of which were almost certainly never built. These included a huge crossbow, various flying machines, a parachute, an armoured vehicle, a dredging machine, a

**ABOVE:** Model of a revolving crane. Leonardo's twin cranes were designed for quarrying. Stones cut from a rock face would be loaded into one bucket; the whole crane would then rotate, and the bucket would be emptied while another was loaded.

**LEFT:** Model based on Leonardo's design for a screw-cutting machine. Turning the crank handle causes the dowel in the centre to turn. At the same time, it turns the two side screws, advancing the cutting tool along the length of the wooden dowel in the centre.

helicopter, a humanoid mechanical robot, an aqualung, a bicycle and a water-powered alarm clock.

In recent years, several of his inventions have, at last, been constructed and work remarkably well, albeit with a bit of adaptation in some cases. A few of Leonardo's inventions did make it out of his notebooks in his lifetime, including machines for bobbin-winding and lens-grinding machine.

In 1513, Leonardo met the king of France, Francis I (1494–1547), after the king's conquest of Milan. Francis commissioned him to make him an automaton in the form of a lion. Leonardo made one that walked, turned its head and even presented a bunch of orchids when stroked. Leonardo died peacefully in France, renowned for his astonishing artistic skill but almost unknown for his scientific insight and his remarkable inventions.

Leonardo imagined fabulous machines that could carry a man through the air, protect humanity from blasts or present bouquets. Other inventors set their minds on more earthy practicalities. Between translating Latin texts and falling out of favour with Her Majesty Queen Elizabeth I, English courtier Sir John Harington (1561–1612), was considering the messy business of human waste. The traditional garderobe of medieval castles was messy and unpleasant, chamber pots and "close-stools" no better, so he set his mind to inventing the precursor of the indoor, flush-toilet. He installed his contraption for the Queen at her palace in Richmond which, given that Harington had failed to also invent trap plumbing (this would have to wait until 1775, when Alexander Cummings invented the S-bend) included a stopper against the worst of any foul odours. Unfortunately, a pamphlet describing his invention, *The Metamorphosis of Ajax*, was also a somewhat unsubtle comment on society. It was written in terms so fruity that Harington was once again banished from court.

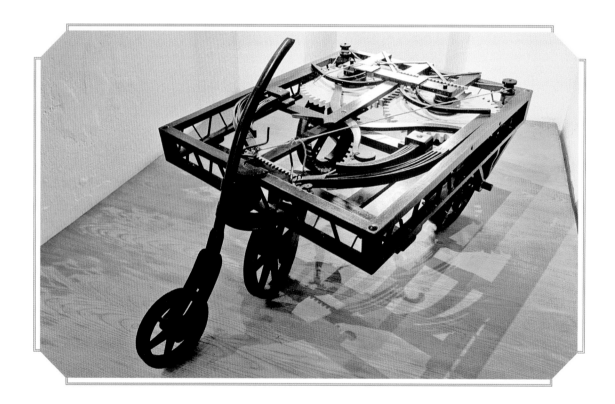

**ABOVE:** Model of Leonardo's car. Leonardo intended it to be powered by spring-driven clockwork. It has no driver's seat, because this was designed to be an automaton. Like most of Leonardo's remarkable inventions, the car was not built in his lifetime.

**OPPOSITE:** Leonardo's assault tank – a model built by IBM and on display at Château du Clos Lucé, France, Leonardo's final home. The shell of this hand-cranked tank was reinforced with metal plates containing holes so that the soldiers could fire weapons from within. Above it can be seen the sketches he made, on which the model was based.

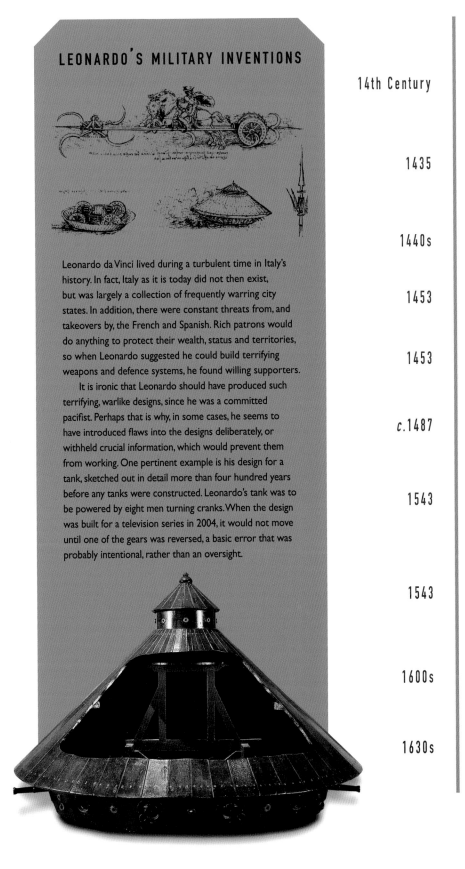

# LEONARDO'S MILITARY INVENTIONS

Leonardo da Vinci lived during a turbulent time in Italy's history. In fact, Italy as it is today did not then exist, but was largely a collection of frequently warring city states. In addition, there were constant threats from, and takeovers by, the French and Spanish. Rich patrons would do anything to protect their wealth, status and territories, so when Leonardo suggested he could build terrifying weapons and defence systems, he found willing supporters.

It is ironic that Leonardo should have produced such terrifying, warlike designs, since he was a committed pacifist. Perhaps that is why, in some cases, he seems to have introduced flaws into the designs deliberately, or withheld crucial information, which would prevent them from working. One pertinent example is his design for a tank, sketched out in detail more than four hundred years before any tanks were constructed. Leonardo's tank was to be powered by eight men turning cranks. When the design was built for a television series in 2004, it would not move until one of the gears was reversed, a basic error that was probably intentional, rather than an oversight.

## TIMELINE

**14th Century** The Renaissance is generally considered to have begun in Florence, Tuscany, and signifies the end of the Middle Ages and the rebirth of the spirit of enquiry of ancient Greece and Rome.

**1435** The incredibly rich banking family the Medici begins its dynasty in Florence. It will go on to fund many of the Renaissance's most important players, including Leonardo, Michelangelo and Galileo.

**1440s** Johannes Gutenberg invents the printing press, allowing new ideas to spread to other parts of Europe.

**1453** The fall of the Byzantine city of Constantinople causes Greek scholars to flee to Italy, taking with them texts rich with classical ideas.

**1453** The end of the Hundred Years' War, between France and England, allowed Renaissance ideas to spread to those countries.

**c.1487** Leonardo produces his drawing *Vitruvian Man*, which typifies the happy union of art and science during the Renaissance.

**1543** Polish astronomer Nicolaus Copernicus publishes *De Revolutionibus Orbium Coelestium* (*On the Revolutions of the Celestial Spheres*), a book that challenges orthodoxy and the church, by proposing that the Sun, and not the Earth, is at the centre of the Universe.

**1543** Flemish physician Andreas Vesalius publishes *De Humani Corporis Fabrica* (*On the Fabric of the Human Body*), which overturns much of the assumed knowledge on human anatomy.

**1600s** New scientific instruments, such as thermometers, microscopes and telescopes, encourage experimentation and observation, leading to new theories and understanding.

**1630s** Italian mathematician Galileo Galilei challenges the accepted understanding of force and motion, leading the way for English scientist Isaac Newton to formulate his laws of motion in 1687.

25

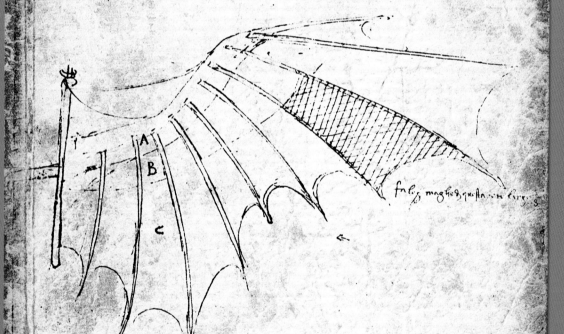

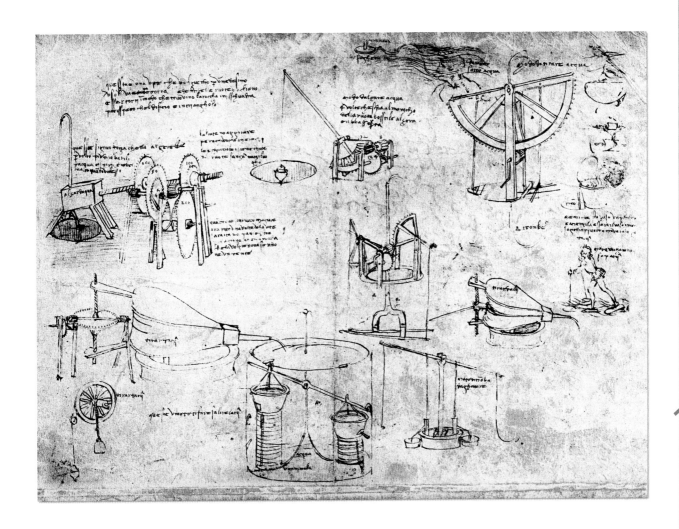

**OPPOSITE:** Leonardo's design for the wings of a flying machine, c.1485, based on his careful observation of bats' wings. Leonardo drew designs for parachutes, helicopters, gliders and human-powered "ornithopters"; there is no conclusive evidence that he ever built and tested his designs.

**ABOVE:** Sketches of various ideas, c.1480 – including a way of walking on water (to the right of the page), using cork "skis" and poles that act like paddles, and a water fountain that pumps water from a well to a goose-neck tap (left).

# Telescopes & Microscopes

## THE TELESCOPE HAS ENABLED US TO DISCOVER OUR PLACE IN THE UNIVERSE, AND TO REVEAL THE TREASURES AND SHEER SCALE OF DEEP SPACE.

Nobody is really sure who was the first to construct a practical telescope or whose genius was the first to realize the potential for this device. But Dutch lens maker Hans Lipperhey (1570–1619) was the first to apply for a patent, in 1608.

Hans Lipperhey (sometimes spelled Lippershey) was born in Wesel, Germany, and moved to Middelburg, in the Netherlands (then the Dutch Republic), in 1594. In the same year he married, became a Dutch citizen and opened a spectacle shop in the city. Little is known of his life, but what is clear is that he was the first person to apply for a patent for the telescope, which was called a "*kijker*" (Dutch for "viewer").

In September 1608, Lipperhey travelled to The Hague, the political centre of the Dutch Republic, where he filed the patent application for his device. His application was denied, because of the simplicity of the invention – it was really just two lenses held at a certain distance apart in a tube. However, the officials at The Hague saw the potential of Lipperhey's instrument, and commissioned him to build three sets of double-telescopes (i.e. binoculars). The Dutch States General paid Lipperhey handsomely for his work; he received more than enough to buy the house next to his and pay to have major renovation work carried out.

**ABOVE:** Compound microscope designed by English scientist Robert Hooke (1635–1703), whose 1665 book, *Micrographia*, revealed the microscopic world to the public for the first time. Unfortunately, Lipperhey died long before the book was published. The glass balls and lenses focused light onto the specimen.

29

**ABOVE:** Artist's impression of Hans Lipperhey in his workshop, experimenting with lenses during his invention of the telescope. The eyepiece lens magnifies the image produced by the larger, objective lens. The lens-grinding machines and lathes are powered by treadles beneath the benches.

**LEFT:** Lens-grinding machine, designed by Leonardo da Vinci. Lipperhey would have used a machine like this to grind concave lenses for the eyepieces of his telescopes and a slightly different machine to make the larger, convex lens that collects the light (the objective lens).

As it turns out, the States General was probably justified in refusing Lipperhey a patent. Within a few weeks, another Dutch spectacle maker, Jacob Metius (1571–1630), submitted a very similar application. In the 1620s, yet another retrospective claim for primacy of the invention of the telescope came to light. Zacharias Janssen (1580–1638), whose house was a few doors away from Lipperhey's, may have beaten Lipperhey to it.

The earliest drawing of a telescope is a sketch in a letter by Italian scholar Giambattista della Porta (1535–1615) in 1609. Della Porta later claimed he had invented the telescope years before Lipperhey, but he died before he could provide evidence of his claim. In fact, it is likely that long before Lipperhey many lens makers had held two lenses in the right configuration and seen a slightly magnified image, but not realized its potential.

Any uncertainty in the story of the telescope falls away in 1609, when other people heard about the new instrument, made their own, and used it for a novel and world-changing purpose: gazing at the night sky. The first person to note that he had gazed upwards in this way was English astronomer and mathematician Thomas Harriot (1560–1621), who made a sketch of the Moon as seen through his telescope on 26 July 1609. Most famously, Galileo Galilei (1564–1642) did the same, and much more, four months later. He published his monumental findings in his book *Sidereus Nuncius* (*The Starry Messenger*) in 1610.

Hans Lipperhey is often also credited with the invention of the microscope, or to be more precise, the compound microscope (consisting of two or more lenses, rather than one). Here again, Zacharias Janssen probably invented the device around the same time as, if not before, Lipperhey. Again, there is no patent for the microscope, because it was inevitable that, at some point, someone would arrange two lenses in the right way to make things look bigger.

Lipperhey's and Janssen's home city of Middelburg was famous for its spectacle makers, thanks to its supply of fine-quality, bubble-free glass and to a superior lens-grinding technique developed in the city. Working with high-quality glass was a novelty in northern Europe in the seventeenth century; the secret of its manufacture had been exported from Italy, which had had the monopoly on fine-quality glass since the thirteenth century.

In a sense, then, along with the lens grinders of Middelburg, the Italian glassmakers of the thirteenth century also deserve credit for these wonderful, world-changing inventions.

### GALILEO GALILEI (1564–1642)

Although Lipperhey was by all accounts a gifted craftsman, and was the first to submit a patent application for the telescope, Galileo is the real genius in this story. His careful and thorough observation of the Moon and his discovery and observations of the moons of Jupiter were key in overturning the longstanding, dogmatic theory that the Earth is at the centre of the Universe.

Galileo improved the basic telescope design, and by August 1609 had managed to make his own instrument with a magnification of 8x (8-to-1), compared to Lipperhey's instrument, which could only magnify 3x. In the 1610s, he also experimented with the compound microscope, and in the 1620s, he became one of the first to make biological observations with microscopes.

Galileo was a great thinker, and is often called the "Father of Physics" or even the "Father of Modern Science". He was much more a pure scientist than an inventor, although he did invent a primitive thermometer and a geometrical compass, and he did not actually invent the telescope.

**OPPOSITE:** The Hubble Space Telescope, in orbit above Earth's atmosphere. Hubble has a concave mirror, rather than an objective lens, to gather light. A camera inside takes pictures using that light, producing incredible, clear images of a wide range of astronomical objects.

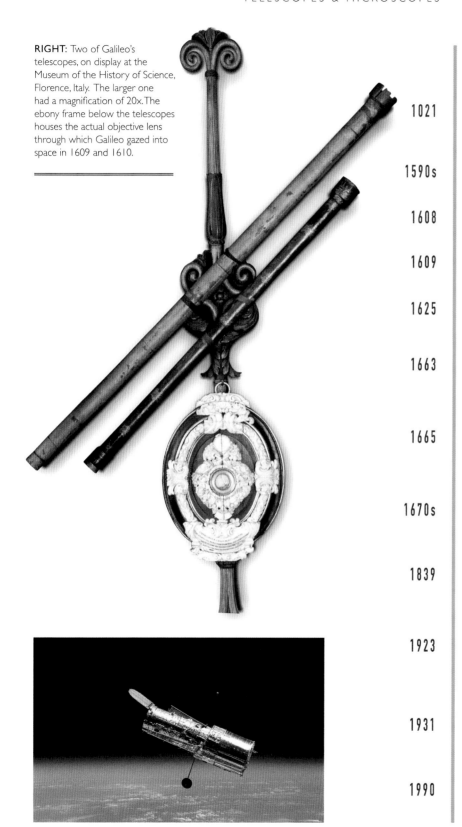

**RIGHT:** Two of Galileo's telescopes, on display at the Museum of the History of Science, Florence, Italy. The larger one had a magnification of 20x. The ebony frame below the telescopes houses the actual objective lens through which Galileo gazed into space in 1609 and 1610.

## TIMELINE

**1021** Islamic scientist Ibn al Haytham, or Alhazen, publishes *Book of Optics*, where he explains how lenses work. Lenses have been used, for magnification and more, since at least 1000 BCE.

**1590s** Dutch spectacle makers Hans and Zacharias Janssen produce a crude microscope in their workshop.

**1608** Hans Lipperhey presents his patent application for the telescope to the Dutch States General.

**1609** Galileo Galilei builds powerful telescopes, and uses them to make groundbreaking astronomical discoveries.

**1625** Italian polymath Francesco Stelluit (1577–1652) becomes the first person to use a microscope for scientific studies.

**1663** Scottish mathematician James Gregory describes the reflecting telescope, which has a concave mirror instead of a lens. "Reflectors" are used by most large observatories today.

**1665** Publication of Robert Hooke's *Micrographia*, which includes spectacular engravings that reveal the details of the microscopic world to a wide audience for the first time.

**1670s** Using his powerful home-made microscope, Dutch amateur Antonie van Leeuwenhoek becomes the first person to observe micro-organisms, which he calls "animalcules".

**1839** The beginning of astronomical photography, as American scientist John Draper takes the first photos of the Moon through a telescope.

**1923** Using the Hooker Telescope at Mount Wilson Observatory, California, with its 2.5-metre (100-inch) diameter mirror, American astronomer Edwin Hubble proves that our galaxy is only one of countless millions.

**1931** German scienists Max Knoll and Ernst Ruska invent the electron microscope, which uses electromagnetic fields to focus beams of electrons, rather than lenses to focus light.

**1990** NASA launches the Hubble Space Telescope, which operates continuously for more than 20 years, advancing our understanding of the Universe.

31

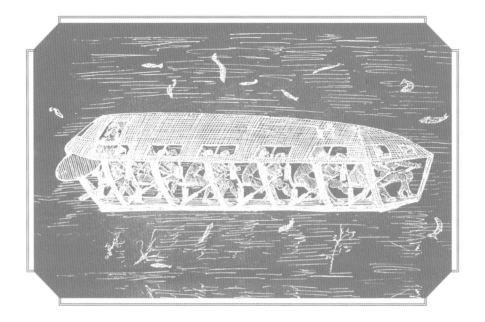

# The Submarine

FOR THOUSANDS OF YEARS THE IDEA THAT HUMANS
COULD SURVIVE UNDERWATER FOR LONG PERIODS
SEEMED AS CRAZY AS SOMEONE FLYING. THIS DID NOT
STOP THE WORLD FROM DREAMING.

Cornelius Drebbel was born in Alkmaar, in the Netherlands (then the Dutch Republic), the son of a wealthy farmer. He had little formal schooling but, aged 20 he was apprenticed to Dutch painter, engraver and publisher Hendrick Goltzius (1558–1617) in Haarlem. Here, Drebbel experimented with the mysteries of alchemy, and throughout the rest of his life, his work, including ingenious inventions, was dominated by the elements of that art: earth, air, fire and water.

From 1604 until his death, Drebbel created many new and improved inventions, including the one that would bring him fame: a fascinating astronomical clock called the Perpetuum Mobile. It was powered by changes in air pressure and temperature, a fact that Drebbel was aware of but, happy encourage a little mystique, he allowed the world to think it could run for decades without a visible source of power. He followed it up with a process for making an intense scarlet dye, a portable bread oven and a thermostatically-controlled furnace (the first known autonomous control system).

Drebbel kept inventing. He came up with a primitive thermometer and an early form of air conditioning, experimented with light and

---

**ABOVE:** According to the "Chronicle of Alkmaar" of C. vander Wreede, van Drebbel built a ship which, driven by 12 oarsmen, is said to have made underwater a journey of several hours at a depth of 12-15 feet. Legend has it that King James even had a ride in the vessel – in which case he would have been the first monarch ever to travel in a submarine.

## JEANNE VILLEPREUX-POWER (1794-1871)

Not exactly a household name today, Jeanne Villepreux-Power was fascinated by underwater marine life, but found it hard to study the creatures in their natural habitat. In the absence of suitable underwater equipment, she invented her own.

Daughter of a country shoemaker, at 18 she walked to Paris to work as a dressmaker's assistant. She proved deft and creative, even designing a royal wedding gown, which brought her enough fame to begin mixing in different circles. She married a wealthy English merchant, James Power, and, at her new home in Sicily, taught herself natural history, studying the island's fish and marine creatures in depth. She was particularly interested in a species of octopus, the paper nautilus *A. argo*. In 1832, in order to study the animals where they lived, she invented the first recognizable glass aquarium. Through her invention, Villepreux-Power was the first to discover that *A. argo* makes its own shell; prior to her research people had assumed it appropriated its shell from other creatures. By now an expert in her field and an accomplished illustrator, in 1832 Villepreux-Power published a groundbreaking book, *Physical Observations and Experiments on Several Marine and Terrestrial Animals*.

Villepreux-Power went on to invent other underwater devices, including a glass vessel inside a cage for shallow-water observation, and one that could send its contents to lower depths. In 1858, British biologist Richard Owen described Villepreux-Owen as "the mother of aquariophily."

lenses, constructed an early form of projector, and engineered one of the first practical microscopes. Even his automatic chicken incubator, however, would pale next to his most famous invention, the world's first submarine.

Sadly, no convincing illustrations of Drebbel's invention exist, but there are contemporary accounts and modern best-guesses of how he might have built it. Between 1620 and 1624, while working for the English Royal Navy, Drebbel built three different versions of his vessel. He tested them in the River Thames in London and eyewitness

**BELOW:** This modern reconstruction of one of Drebbel's submarines sits in Heron Square, Richmond-upon-Thames, London. It was based on design documents found at the Public Records Office in London, and made by a local boatbuilder in 2003.

accounts suggest that the craft could stay submerged for hours at a time, diving four to five metres (13–16 feet) beneath the surface.

Drebbel's submarines were sealed wooden double-hull vessels with leather-sealed holes along the sides through which oars protruded. They were covered with greased leather to make them watertight and contained large pigskin bladders for buoyancy, filled with, and emptied of, water as necessary. The third and largest vessel could carry 16 people, 12 of them oarsmen. Some accounts suggest that long tubes allowed the oarsmen to breathe, but there is also evidence that Drebbel may have used a chemical reaction – heating saltpetre (potassium nitrate) – to produce oxygen.

Drebbel tried to convince the English Royal Navy to adopt his submarine for use in warfare but, despite his ongoing relationship with the royal family, the Navy was not interested. It would be 150 years before submarines were used for military purposes.

An inventor who had only slightly more luck with the military was David Bushnell (1742–1824). Born into a farming family, Bushnell decided at the age of 31 to study mathematics at Yale. He graduated in 1775, at the outbreak of the American Revolution. As a fierce patriot, Bushnell wanted passionately to help the cause by inventing something useful. His underwater bomb, the forerunner of

modern naval mines, needed some method of being safely attached to an enemy warship. He created a strange, barrel-shaped vessel, called *Turtle*, large enough to take a man, that could fully submerge and move underwater by the operator turning two propellers by hand. A pair of inventor friends, Phineas Pratt and Isaac Doolittle, came up with a clockwork detonator that would delay the striking of a musket flint, allowing the operator to get away before the mine exploded.

Bushnell's experiment ultimately failed, not because the Turtle wouldn't work, but because the only person who understood how it worked, Bushnell himself, wasn't healthy or strong enough to take it into active service. The *Turtle* was eventually lost during the Battle of Fort Lee, but Bushnell went on to develop more successful underwater mines. The propellers he developed for the *Turtle*, based on the way marine animals move, have inspired submarine design ever since.

**LEFT:** Contemporary portrait of Dutch inventor Cornelius Drebbel from an engraving by an unknown artist, published in 1628.

**ABOVE:** Illustration of Drebbel's Perpetuum Mobile clock, from the 1612 book *A Dialogue Philosophicall* by English clergyman and author Thomas Tymme (d.1620). The central sphere (A) represents the Earth, while the upper sphere (B) displays the lunar phases.

ABOVE: Model of *Turtle*, the first submarine to be used in warfare. The one-man craft, driven by hand-cranked propellers, was built in 1775 by American inventor David Bushnell (1742–1824), and was used to attach explosives to the hulls of ships.

## TIMELINE

**1578**
English mathematician William Bourne comes up with the first design for a submarine. In some ways, it was more advanced than Drebbel's, as it included buoyancy tanks – but it was never built.

**1620s**
Cornelius Drebbel builds the first successful submarines; he tests three designs, each larger than the last.

**1775**
American inventor David Bushnell designs and builds *Turtle* – the first submarine to be used in warfare.

**1834**
Russian engineer Karl Schilder designs the first all-metal submarine.

**1854**
French inventor Marie Davey designs the first periscope to be used in submarines. The first collapsible periscope was designed by American naval architect Simon Lake.

**1863**
The French vessel *Plongeur* becomes the first submarine to be driven mechanically, rather than by human power. It is powered by a compressed air engine.

**1896**
Irish-born American naval engineer John Holland designs the Holland Type VI – the first submarine to be powered by a petrol engine when at the surface and by electric motors when submerged.

**1914–18**
The German navy uses a total of 360 U-boats to sink more than 11 million tonnes of shipping.

**1954**
The US Navy's *Nautilus* is the first nuclear-powered submarine. In 1958, the vessel reached the North Pole, travelling under the ice.

**1960**
French aquanaut Jacques Piccard and US Navy Lieutenant Don Walsh reach the deepest part of the ocean – the Challenger Deep – in a bathyscaphe called *Trieste*.

# The Age of Electricity

## THE UNITED STATES OF AMERICA WAS BORN ON 4 JULY 1776; AT THE SAME TIME, THE IDEA OF HARNESSING ELECTRICITY WAS BEING INTRODUCED TO THE WORLD.

A man involved in both of these endeavours was statesman Benjamin Franklin (1706–1790), an important figure in eighteenth-century science and invention. He was born in Boston, Massachusetts, USA, one of 17 children, and his parents could only afford to send him to school for two years. He was keen to learn, however, and was an avid reader – and at just 12 years old, he became an apprentice at his older brother's printing firm. Following a dispute with his brother five years later, Franklin ran away to make a new life in Philadelphia. Penniless, he managed to find an apprenticeship in a printer's firm there and soon set up his own printing shop.

By the 1740s, Franklin was very successful. He strongly believed that science and technology could be used to improve society and, in 1743, founded the American Philosophical Society, the nation's first learned society. In the same year, he invented a cleaner, more efficient way of heating the home: the Franklin Stove. Since he intended it to be for the public good, he didn't patent it.

In 1749, Franklin retired from business, to spend more time on his research. His work on optics famously led him

**ABOVE:** New York's Empire State Building being struck by lightning. At the very top of the building is a lightning rod that helps to discharge thunderclouds over the city. In addition to draining charge away, the building receives about 100 lightning strikes each year.

to invent bifocals, although others probably invented them independently, around the same time.

Fire prevention was a major concern at the time, since most buildings were still made of wood. In 1736, Franklin had founded one of America's first volunteer fire departments. In 1752, he formed America's first fire insurance company, and came up with his most famous invention, the lightning rod, aimed at preventing the risk of fire from lightning.

Lightning rods, or lightning conductors, are pointed metal spikes connected to the earth, which draw off electric charge from clouds, dramatically reducing the risk of lightning strikes. When lightning does strike, the rods carry the electricity to the ground, bypassing the building to which they are attached. They may seem like a simple or even insignificant invention today, but at the time, Franklin's invention caused a real buzz and helped to foster the idea that basic insight into natural forces can produce important practical results.

Franklin's fascination with electricity and lightning led him to carry out his famous kite experiment, in 1752. During a storm, he flew a kite into a thundercloud and drew electric

**BOTTOM:** Coloured lithograph illustrating Franklin's 1752 experiment that proved lightning is an electrical phenomenon. On the ground, beside Franklin, is a Leyden jar to collect electric charge drawn off the thundercloud through the kite string.

**BELOW:** Eighteenth-century Franklin-style bifocals, with sliding adjustable arms. In a letter to his friend, English merchant George Whatley, dated 1784, Franklin wrote that he was "happy in the invention of double spectacles", although it is possible someone else had invented them before him.

37

38

**LEFT:** A Voltaic Pile, the world's first battery. Batteries were invented shortly after the death of Benjamin Franklin, but their name is derived from his work, after he compared the effect of his Leyden jars to a "battery of cannons".

**BELOW:** A modern 1.5 volt cell, equivalent to one layer of Volta's battery. The pile of cells in Volta's battery was 30 centimetres tall alone, compared to the below batteries which are around 5 centimetres in length, but have more power and last longer than Volta's entire battery.

charge down the wet kite string, proving for the first time that lightning is an electrical phenomenon.

Even today, the world runs on batteries. Alessandro Volta (1745–1827) was the first to realize where that power originates. Volta's school wanted him to be a priest. His family wanted him to take law as a profession. Volta himself was fascinated by physics. He began, aged 18, to exchange letters and ideas with leading physicists of the day.

In 1791, Volta's friend Luigi Galvani (1737–98) demonstrated an experiment where the muscles of a frog, stretched between two pieces of metal, generated an electric current. He announced to the world he had discovered "animal electricity". Volta wasn't so sure. He thought the frog was probably just a conduit between two pieces of metal, creating "metallic energy". There were advocates for both schools of thought but Volta, having tested pieces of metal on his tongue and detecting a small current, was able to prove his point in 1800 with the world's first electric battery.

The Voltaic Pile, or Column, the first battery, was a pillar of alternating metal discs of zinc and silver (or copper and pewter) separated by cloth soaked in salt water or sodium hydroxide, able to continuously provide electric current to a circuit.

## RESEARCHES IN ELECTRICITY

Franklin conceived of the lightning rod after carrying out researches into electricity, a hot topic at the time. In 1747, he set up a laboratory at his own home. In the mid-1740s, scientists in Germany and Holland had invented a way of storing large amounts of electric charge, in a device called a Leyden jar (below). Franklin connected several of these jars together, so that they could produce a much stronger effect.

In five ground-breaking letters to Britain's Royal Society, Franklin laid down the foundations for the proper study of electrical phenomena. He was the first person to use the terms "charge" and "discharge", the first to write about "positive" and "negative" electricity, and the first to understand that electric charge is not "created", but simply transferred from place to place.

## TIMELINE

**c.600 BCE** — Greek philosopher Thales of Miletus (c.624–c.546 BCE) is the first to write about electrically charged materials – especially amber, which picks up light objects when rubbed.

**1600** — English physician William Gilbert (1544–1603) invents the term "electricity" based on the Greek word for amber (*elektron*).

**1660** — German scientist Otto von Guericke (1602–1686) invents an electrostatic generator, allowing experimenters to create large amounts of electricity at will.

**1740s** — Scientists in Leiden, then in the Dutch Republic, invent the Leyden jar – a device to store large amounts of electric charge.

**1747** — Franklin sends the first of a series of letters to English scholar Peter Collinson (1694–1768), in which he establishes the standard terminology for electricity research, including "positive" and "negative".

**1752** — Franklin flies a kite in a thunderstorm, proving that lightning is an electrical phenomenon.

**1799** — Italian scientist Alessandro Volta (1745–1827) invents the battery – a pile of alternating copper, zinc and brine-soaked cardboard discs.

**1827** — German physicist Georg Ohm (1789–1854) publishes Ohm's Law, defining electrical resistance as the mathematical relationship between current and voltage.

**1859** — French physicist Gaston Planté (1834–1889) invents the first rechargeable battery.

**1899** — English physicist Joseph John "JJ" Thomson (1856–1940) discovers the electron, the elementary particle responsible for most electrical effects.

The scientific world was agog. Within six weeks, two English scientists, William Nicholson (1753–1815) and Anthony Carlisle (1768–1840) had used one of Volta's devices to decompose water into hydrogen and oxygen, instigating a whole new field of scientific research – electrochemistry.

Volta himself was showered with honours and is remembered today in our term "volt" – a unit of the electromotive force that drives current.

# The Cotton Gin

IN 1787, VIRTUALLY NO COTTON WAS BEING GROWN IN THE
US. IT TOOK THE INVENTION OF THE COTTON GIN TO
TURN IT INTO THE CROP THAT REVOLUTIONIZED THE
AMERICAN ECONOMY.

By speeding up how long it took to remove husks and
sticky seeds from raw cotton fibre, the cotton gin ("gin" being a
contraction of "engine") transformed the production of cotton in
the US. The machine itself was straightforward enough. It used
wire teeth hammered into a rotating wooden cylinder to snare the
raw cotton fibres and force them through a wire mesh, which had
holes small enough to trap the husks and seeds. The question is,
who actually invented the gin?

Traditionally, sole credit for the invention is given to
Connecticut-born teacher Eli Whitney (1765–1825), who
patented the original cotton gin in 1794. However, recent
historians differ. Some say that it was Catharine Littlefield
Greene (1755–1814) – widow of Nathaniel Greene, a leading
American Revolutionary War general and owner of Mulberry
Grove, a plantation near Savannah – who actually devised the
machine and that Whitney, who was tutoring her children at the
time, simply built it by following her drawings. Others believe
that the gin was Whitney's idea, and Greene was the person who

**ABOVE:** A model of Greene
and Whitney's original cotton
gin, which is on display at the
Smithsonian Institution in the
US. The model is said to have
been made before 1800 and was
deposited to the museum by Eli
Whitney, Jr., in 1884.

actually made it practical by suggesting the substitution of wire teeth for the ineffective wooden ones Whitney had originally specified. Still more hold that the invention was the result of a collaboration between Whitney, Greene and a slave labourer on her plantation whose name remains unknown.

What is generally accepted is that Greene provided the money Whitney and Phineas Miller, her second husband and Whitney's partner, needed to finance the manufacture of the gin. She also helped with its marketing to neighbouring plantations, though the way in which this was done proved to be a mistake. The aim was not to sell gins outright, but to lease them to the plantation owners in exchange for a percentage of the profits their use would create. Many plantation owners balked at this notion and, because the gin was relatively easy to make, simply copied it. Soon, bootleg imitations were being put into use throughout the South. Greene spent much of her fortune fighting court battles in a vain attempt to protect the patent Whitney had taken out.

Just four years later, the company Whitney and Miller had founded to manufacture the gin was forced into bankruptcy. As Whitney ruefully noted, the device proved invaluable to everyone but its inventor. As to why Greene did not try to patent the gin herself, it is likely that, despite being considered unconventional by many of her peers, she realized that attempting to do so would have exposed her to ridicule and robbed her of her position in Southern society. She simply stood by and allowed Whitney to take the credit.

## SEWING THREAD AND SEWING MACHINES

Unlike Littlefield Greene, fellow American inventor Hannah Wilkinson Slater (1774–1812) was not shy in applying for her own patent for "a new method of producing sewing thread from cotton". The patent was issued in 1793; it was the first ever to be granted to a woman.

A native of Pawtucket, Rhode Island, Hannah was only 16 when she married Samuel Slater, a young British entrepreneur who had emigrated to the US to set up his own cotton mill. It was Slater who presented his wife with some smooth yarn he had spun from Surinam cotton, but it was she who realized the yarn's possibilities. Working with her sister, she hand-spun it into a new and stronger type of two-ply thread that was a great improvement on the linen thread previously used for sewing cloth. Hannah took out her patent the year she made her discovery.

Hannah Slater was relatively short-lived; she died as a result of complications following childbirth at the age of 37. Helen Augusta Blanchard (1840–1922) lived for far longer. She was a prolific inventor with no fewer than 28 patents issued in her name. Her ingenuity helped to transform the textile, hat and corset industries.

Born in Portland, Maine, the daughter of a wealthy shipowner, Blanchard began inventing seriously only when she was already over 30, after the family business had gone bankrupt, a casualty of the great financial crash of 1866.

Her most celebrated invention was the zigzag sewing machine, which she patented in 1873. The zigzag stitch the machine produced sealed the raw edges of a seam, making the resulting garment sturdier. It made possible the automatic production of a host of dry goods, whose manufacture had previously depended on time-consuming hand stitching.

41

**LEFT:** A wood engraving of an original cotton gin that shows some of the machine's inner workings. The handle turned the central cylinder, which was covered in sharp wire teeth. The teeth separated the hard cotton seeds from the soft cotton fibre, which was then collected for processing.

# Steam Power

## ONE AFTERNOON IN MAY 1765, SCOTTISH ENGINEER JAMES WATT HAD AN IDEA THAT CHANGED THE WORLD.

Watt (1736–1819) had hit upon a clever device to make steam engines more efficient and more powerful. It was this device and his other inventions that made steam the driving force of the Industrial Revolution.

English engineer Thomas Newcomen (1663–1729) built the first practical steam engine in 1712 to pump water from coal mines. By the time of Watt's birth, there were nearly a hundred Newcomen engines across Britain, and several more in other countries.

Newcomen's engine relied on atmospheric pressure to push down a piston inside a huge, open-topped vertical cylinder. That could only happen if there was a vacuum inside the cylinder, beneath the piston. Newcomen achieved the necessary vacuum by condensing the steam inside the cylinder back into water, which takes up only a tiny fraction of the volume steam does. A system of valves allowed steam to fill the cylinder, then sprayed in cold water to condense the steam. Having to cool the cylinder

down for each stroke of the piston, and then heat it up with steam ready for the next stroke, made the engine incredibly inefficient. It was this fact that Watt addressed that day in 1765.

James Watt was born in Greenock, a town on the River Clyde, west of Glasgow in Scotland. His father was a ship's

**ABOVE:** Watt rotative engine at the Science Museum, London. In the background is the cylinder; to the right, the speed-regulating governor (invented by Watt); in the foreground, the flywheel.

**OPPOSITE:** Reconstruction of Watt's workshop at the Science Museum, London, after the contents were removed from Heathfield Hall, Handsworth, in Birmingham. Watt was using the busts on the workbench to test a machine he invented to copy sculptures – a kind of three-dimensional photocopier.

## HENRIETTA VANSITTART (1833–1883)

When nineteenth century inventor James Lowe (1798–1866) was tragically killed in a street accident, his fourth child Henrietta was determined to continue his work. The screw propeller Lowe had invented for ships, using curved blades, fascinated Henrietta and, in 1857, she had accompanied her father to HMS *Bullfinch* to test out his latest version. She thought it could be improved and, after her father's death, without any scientific or engineering training, she began work on modifications. Two years later the Lowe-Vansittart propeller was patented and in use on HMS *Druid*, fitted by Henrietta herself. Ships could now move faster and with less fuel; the propeller would be used on many new vessels, including the ill-fated *Lusitania*. Henrietta was lauded by the Admiralty, won prizes and was mentioned by name in the *Times*.

She had married Lieutenant Frederick Vansittart in 1855, but had a long-lasting affair with Edward Bulwer-Lytton, a cabinet secretary in Disraeli's government. Bulwer-Lytton died in 1873, and Henrietta was beginning to show signs of mental illness. She continued with her work, becoming, it is believed, the only woman to write, illustrate and present her own scientific articles before the members of a scientific institution, the Association of Foreman Engineers and Draughtsmen.

In 1882, she was found wandering and confused. She was committed to a lunatic asylum in Gosforth, near Newcastle on Tyne, a long way from her London roots, and died in 1883 of acute mania and anthrax. The journal of the London Association of Foreman Engineers and Draughtsmen described her in a glowing obituary as having "a great knowledge of engineering matters and considerable versatility of talent".

instrument builder. Using a tool kit his father had given him, Watt became a skilled craftsman from an early age. Following a year working in Glasgow, and a year in London learning the trade of making mathematical instruments such as theodolites and compasses, Watt wanted to set up his own shop. After repairing an instrument for a professor at Glasgow University, he was offered a room there to use as a workshop, and earned a living making and selling musical instruments as well as mathematical ones.

In 1763, Watt began experimenting with a model of a Newcomen engine. He quickly realized just how much fuel, steam and heat Newcomen's design wasted. Watt's great idea of 1765 was the "separate condenser". In Watt's design, the steam was condensed in a chamber connected to but separate from the cylinder. The chamber was held at a lower temperature, so that the cylinder could remain at boiling point. Watt patented his invention in 1769. The engineer and entrepreneur Matthew Boulton (1728–1809) went into business with Watt in 1775. Their partnership lasted until Watt's retirement in 1800 and completely revolutionized the use of steam engines in industry.

Until 1782, steam engines were still used only to pump water in coal mines. That year, on Boulton's request, Watt invented a way to make a steam engine produce a rotary motion, rather than an up-and-down motion – and the resulting "rotative" steam engines were an immediate success. Before long, Watt's rotative engines were installed in textile mills, iron foundries, flour mills, breweries and paper mills.

Watt made many other important improvements to steam power – including, in 1782, the "double-acting" engine where steam was admitted to the cylinder alternately above and below the piston, all choreographed by a clever system of automatic valves. He also invented a steam-pressure gauge and a way of measuring the efficiency of a steam engine. In 1788, he invented the "governor", a device that automatically regulated the speed of an engine.

Watt was also a respected civil engineer, working mostly on canal projects. He is credited with other inventions too, including a popular device for making multiple copies of letters. But steam was his life's work. In honour of his achievements in steam power, the international unit of power, the "watt", is named after him.

**RIGHT:** Portrait of James Watt, painted around 1810. Watt's inventions revolutionized the use of steam power in industry in the 1770s and 1780s.

## THE LUNAR SOCIETY

James Watt was a member of a very important society: an informal group of scientists, engineers, industrialists, philosophers, doctors, artists and poets called the Lunar Society.

This group of intellectuals typified the spirit of the Age of Enlightenment – that period of history when people began believing that science, technology and reason could, and should, shape society. Their activities centred on regular meetings, which were often held at the house of Matthew Boulton, also a member. In addition to the meetings, the members of the group were in frequent communication by letter.

The Lunar Society was very important in the transformation of Britain from a rural, agricultural society to an urban, industrial one – it has been described as the revolutionary committee of the Industrial Revolution. The society's name was derived from the fact that the meetings were always held on the Monday closest to Full Moon; the moonlight made it easier for members to get home afterwards.

**OPPOSITE:** Model of an early Savery pumping engine. Steam from the boiler (left) filled the receiver (right), and water rushed into the receiver as the steam condensed. The water was then forced up through another pipe when new steam was admitted to the cylinder.

## TIMELINE

**100** Greek mathematician Hero of Alexandria (c.10–70 CE) invents the aeolipile, a spinning ball powered by jets of steam. Today it would be called a reaction turbine.

**1601** Italian scholar Giambattista della Porta (c.1535–1615) experiments with steam, producing partial vacuums as steam condenses.

**1606** Spanish soldier Jerónimo de Ayanz Beaumont patents a steam-powered pump for draining mines. His design was forgotten after his death (1613).

**1690** French physicist Denis Papin experiments with the vacuum produced by condensing steam, and builds a rudimentary steam engine with a piston moving in a cylinder.

**1698** English inventor Thomas Savery (c.1650–1715) patents the first working steam pump, which he called the "Miner's Friend". It was designed to pump water from mines but was very inefficient.

**1710** English engineer Thomas Newcomen invents the first practical steam engine, which pumps water from mines. The first one was installed in 1712, in Dudley, England.

**1765** James Watt invents the separate condenser, which allows the cylinder to stay hot and brings a massive improvement in fuel efficiency.

**1781** Scottish engineer William Murdoch, who is Watt's assistant, invents the sun and planet gear, which enables Watt's engines to produce rotary motion.

**1782** Watt and his partner Matthew Boulton begin producing rotative engines, which are used to drive machines in factories. English engineer Jonathan Hornblower invents a double-acting steam engine, in which steam acts on each side of the piston in turn.

**1796** English engineer Richard Trevithick (1771–1833) begins work on engines using "strong steam" – steam at high pressure.

**1804** Richard Trevithick's *Penydarren* becomes the first successful steam railway locomotive.

46

*Erste Watt'sche Dampfmaschine. M.~1:20.*   *Disposition für die Aufstellung der Nachbildung in der Hauptwerkstätte München.*

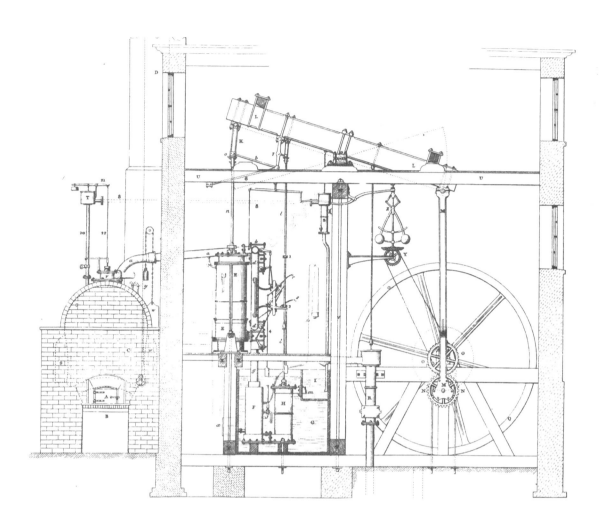

47

**LEFT:** Side and end elevations of Watt's "Lap" engine, 1788. This engine drove lapping (metal-polishing) machines that were previously driven by horses, so when Watt calculated what the machine was capable of, he devised the term "horsepower".

**ABOVE:** Technical drawing of a Watt rotative engine from the 1780s. Steam produced in the boiler (left) entered the cylinder, in which the piston moved up and down. On each down stroke, the piston rod pulled down one end of the pivoting beam, the other end of which turned the large flywheel by means of a sun and planet gear.

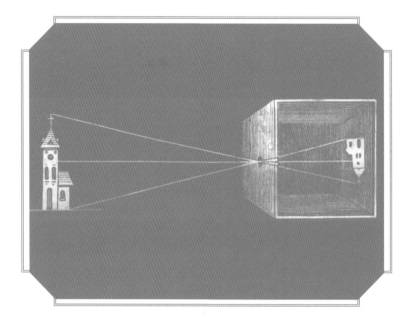

# Photography

LESS THAN 200 YEARS AGO, THERE WAS PRACTICALLY
NO WAY OF PRODUCING A LASTING IMAGE OF A SCENE
OTHER THAN BY DRAWING OR PAINTING IT.

Photography, invented by French scientist Nicéphore Niépce (1765–1833), has had a profound effect on art, education, history and science. Nicéphore Niépce was born in Chalon-sur-Saône, France. His father was a steward to a duke, but little else is known of his childhood. When he was 21, he left home to study at a Catholic oratory school in Angers, where he became interested in physics and chemistry. His first name was originally Joseph; he began using the name Nicéphore, which means "victory-bearer", when he joined the fight against the monarchy in the French Revolution in 1788.

It was in 1793 that Niépce first had the idea of producing permanent images. Around the same time, he and his brother, Claude (1763-1828), conceived of a new type of engine that would use explosions inside a cylinder to drive a piston. Together, they invented the world's first internal combustion

engine, the Pyréolophore. Its fuel was a highly flammable powder of spores from a fungus called lycopodium (which, quite coincidentally, was later used in photographic flash bulbs). They received a patent in 1807, and two years later the brothers entered a government competition to design a replacement for a huge pumping machine on the River Seine in Paris. Their ingenious idea was highly favoured by the judging committee, but in the end the pumping machine was never replaced.

Shortly after its invention in 1796, Niépce learned about a new method of printing illustrations, called lithography, which allowed artists to draw their design directly onto a printing plate, rather than having to etch it into wood or metal. Niépce couldn't draw, so he decided to try and project an image onto the plate instead, hoping to find a way to make the image permanent. To project the image, he turned to an existing technology called the "camera obscura". Popular with Renaissance artists who wanted to produce an accurate representation of a scene, the camera obscura – literally "darkened chamber" – is a simple closed box or room in which a lens casts an image on a screen.

Niépce had some success with paper coated with light-sensitive compounds of silver. Images did register on the paper, but they completely blackened when they were exposed to light as they were removed from the camera. Also, this process produced negatives: the parts of the paper where the most light fell became the darkest parts of the resulting image. So Niépce tried using compounds that bleach in sunlight, instead of those that darken. In 1822, Niépce turned to a substance called bitumen of Judea, a thick, tarry substance that hardens and bleaches when exposed to light. His first real successes were in producing permanently etched metal plates. For this, he placed drawings on top of a sheet of glass, which in turn lay on the metal plate coated with bitumen. After exposure to light, for days at a time, he washed away the unhardened bitumen, then treated the plate with nitric acid. The acid etched into the metal wherever the bitumen was not present, leaving a plate from which he could make prints.

Three years later, Niépce began taking pictures of scenes, rather than "photocopying" drawings. He dissolved bitumen in lavender oil and applied the mixture to pewter plates. Then he exposed the plates for several hours in his camera obscura. The bitumen bleached and hardened where light fell, while the unexposed bitumen – representing the darkest parts of the image – was washed away to reveal the dark metal below. These photographs were not negative, but positive, images. The oldest photograph still in existence is *View from the Window at le Gras* (1826), an eerie image of outbuildings taken from the first floor of Niépce's house.

While people like Louis Daguerre (1787–1851) and William Henry Fox Talbot (1800–1877) were working on straight photography, others were interested in the ways the new processes could be stretched to create images of a different kind.

English botanist Anna Atkins (1799–1871) was keen to find a method of recording specimens in as scientific a way as possible and began a correspondence with Fox Talbot and astronomer/chemist Sir John Herschel (1792–1871). Fox Talbot

**RIGHT:** An 1825 copy of an earlier print. Niépce soaked the print in varnish to make it translucent, then laid it on a copper plate coated with his bitumen solution. After washing the plate in acid, he was left with an etching, from which to make this print.

**LEFT:** In 1826, Niépce dissolved bitumen of Judea in lavender oil, spread it on a polished pewter plate, and exposed it in his camera obscura for about eight hours. The result is this, *View from the Window at le Gras*, the oldest photograph in existence.

**OPPOSITE TOP:** *Table Servie* (*Set Table*) by Niépce. Some experts believe this to be the oldest photograph, dating it to 1822, but it is more likely to be from *c.*1832. The original was on a glass slide, now broken.

**OPPOSITE BELOW:** *The Ladder*, photographed *c.*1845 by William Fox Talbot. Fox Talbot invented the calotype process, which involves making prints from negatives.

## LOUIS DAGUERRE (1787–1851)

After his initial successes with bitumen on pewter plates, Niépce found a way to give better definition to his photographs, or "heliographs" as he called them. He used iodine vapour to make the pewter darken. In 1829, Niépce began collaborating with a French artist, Louis Daguerre. Niépce died in 1833, but by 1837, Daguerre was producing images that only needed a few minutes' exposure. He used copper plates coated with silver iodide, which were "developed" after exposure to mercury vapour and then "fixed" using a strong salt solution.

Daguerre had improved the process so much that he felt justified in calling his photographs daguerreotypes. In 1839, the French Government gave Daguerre's process away, patent-free, as a "gift to the world", and paid Daguerre and Niépce's son a handsome pension. Daguerreotypes became very fashionable, dominating early photography and spurring the development of subsequent photographic technologies.

described his "photogenic drawings" and "calotype" process, using paper coated with silver iodide. Herschel had devised a "cyanotype" process, which produces an image by "sun printing" an item; this process involved laying the item on a piece of paper impregnated with ferric ammonium citrate and potassium ferricyanide and exposing it to light.

Atkins eventually used the cyanotype process to create photograms of dried algae, publishing her work in 1843, in *British Algae: Cyanotype Impressions*. With handwritten text, also created using cyanotype instead of the traditional letterpress print, the work is said to be the first book to use photographic illustration.

**AD 900s**

Arab scientist Ibn al-Haytham (known as; 965–c.1035) conducts the first systematic experiments with the pinhole camera and the camera obscura.

**1400s**

Some Renaissance artists trace camera obscura images to produce a new level of realism.

**1816**

Niépce produces the first photographic images – negatives that soon blackened.

**1823**

Niépce produces photographic copies of prints, using his process called "photogravure".

**1826**

Niépce produces his first permanent photographic images using his camera obscura.

**1832**

French-Brazilian inventor Hércules Florence (1804–1879) independently invents a process very similar to Daguerre's, but remains unknown because he lives in a remote part of Brazil.

**1839**

Having collaborated with Niépce before Niépce's death in 1833, Louis Daguerre perfects his "daguerreotype" process, using light-sensitive silver iodide.

**1841**

English inventor and photographer William Henry Fox Talbot perfects his calotype process, the precursor of modern film photography, in which a negative image is formed on light-sensitive paper. Any number of positive prints could be made from a single negative.

**1848**

English inventor Frederick Scott Archer (1813–1857) introduces the wet plate process. In Archer's process, glass plates were coated with a solution called collodion, in which he dissolved light-sensitive compounds.

**1861**

Scottish mathematical physicist James Clerk Maxwell (1831–1879) produces the first colour photograph, by projecting a red, green and blue image of a tartan scarf.

**1871**

English inventor Richard Leach Maddox introduces the dry plate process, using gelatin instead of collodion.

51

**BELOW / OPPOSITE**: A framed heliograph (photograph) on a pewter plate, produced by Niépce in 1827. The image, of a ruined abbey, is not easy to see in a copy like this: when looking at the actual object, the image appears clear only when it is viewed from an angle.

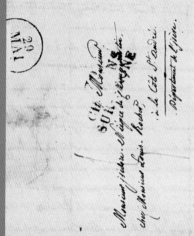

54

**ABOVE / OPPOSITE:** Letter from Niépce to his son Isidore and his daughter-in-law, dated 26 May 1826. Niépce mentions his experiments with heliography — and that he had ordered some sheets of tin. He was referring to pewter, which is an alloy consisting of more than 90 per cent tin.

**TRANSLATION:** Au Gras, 26th May, 1826

We had been waiting very impatiently for news from you, my dear children, when we received your letter yesterday evening, dated 21st May, Trinity Sunday. But it was too late for me to reply by this morning's post. So my reply will not be posted until tomorrow and won't go until Sunday. We were very pleased to learn that your little trip to Nice had been so happy and agreeable. The details you gave us on everything connected with it really interested us, although my wife in particular was very much affected by the loss of her cousin, the only relative she had in that region, and of the other people she knew whom you referred to. I'm afraid that's just what happens in life. It's one of the hardest things we have to go through, but it must be endured, and we in our turn must inflict it on others (but as late as possible!). My wife is very touched, and very grateful, my dear children, by your delicate attentions in connection with a situation which was bound to cause her distress. She was pleased to receive some flowers from her homeland — they, at least, will bring only happy memories. As it happens, we received the two pretty blackcurrant plants almost at the same time as your letter. The post wagon stopped to deliver them here. They were properly packed in straw and had been well looked after, and they were planted straightway. We are delighted with the kindness you received from the Martin family, which came as no surprise to us, as they have always behaved very decently to us. Your unexpected visit must have been a surprise to the family, especially to one of the sons…Isidore knows just what I mean. We think the area around Nice would have seemed more agreeable to you if you hadn't seen Marseille and its surrounding region first, so much is it the case that comparisons often spoil everything. And it must also be admitted that this latter city is much bigger and more beautiful, and has infinitely more to offer. And then, you've got that fellow Ouzet who – well, I won't say he's one of the great and good, but he certainly puts on a great show, which beats all you could ask for. Just look at the enormous block of coral which he has just given to his compatriot. That place is really something overwhelming, and indeed, it's quite difficult to leave that area down there without occasionally being embarrassed or without being ripped off. But we'll talk about that in greater detail when you return.

So far we've received only the foodstuffs you sent, and the 3 boxes sent to Mr Larrier. The 2 barrels of olive oil should have arrived long ago, and the other boxes you told us about some time before you left Marseille should possibly also have arrived here. Remember, I did tell you about this delay in my previous letters. No doubt you took advantage of this, my dear children, to gather information regarding this subject on the spot. This is very much to be desired since, once in Grenoble, you'll no longer have the same opportunities. As for us, every time we have to send to town, we make sure that enquiries are made at Mr Larrier's, at Mrs Baillet's, and at the coach office to check whether anything has arrived.

Oodles of love and affection from your kind relatives at Châlon. They are in excellent health and, above all, they've been very busy these last few days with their move. We still have nothing to report from London. I am spending a lot of time on my heliographic work. I've had some new sheets of tin ordered. This metal serves my purpose better, mainly as regards its nature, as it reflects the light better, so that the picture appears a lot sharper. So I congratulate myself on this happy inspiration.

Antoine and his brother ask to be remembered to you. They learned from Hochamure's son that you were making a trip – he being the only person we told about it, when he came to see us, and actually he knew already. There was also a rumour going round that you'd gone to England. Some people are still stubbornly sticking to this idea. When you're in Grenoble, please don't forget the half-dozen skins of goat kids, which are used for making ladies' gloves, for my pads: I've almost run out of that sort of thing.

My wife joins with me, my dear children, in sending all our kindest thoughts to you.

P. S. All the very best from your friends and relations.

P. S. Please pass on our greetings to the new relatives you're going to visit on the coast. Goodbye!

# The Railways

## DURING THE NINETEENTH CENTURY, THE RAILWAYS COMPLETELY REVOLUTIONIZED TRAVEL AND COMMUNICATION FOR MILLIONS OF PEOPLE.

The coming of the railways was made possible by the invention of high-pressure steam engines by English engineer Richard Trevithick, who also designed and built the first steam locomotives.

Richard Trevithick was born in the parish of Illogan, in Cornwall, England. Richard's father was the manager of several local mines, and young Richard spent much of his early life gaining practical knowledge of steam engines. He did not do well at school, but he earned an excellent reputation after he became a mine engineer, aged 19.

At that time, working engines used steam only at atmospheric pressure or slightly above. Trevithick realized early on that steam under high pressure could lead to more compact, more powerful engines. Most people at the time, including

steam pioneer James Watt (see page 40), feared "strong steam", believing that the risks of explosion were too high. Trevithick began experimenting with high-pressure steam in the 1790s, and by 1794, he had built his first boiler designed to withstand high pressures, from heavy cast iron.

In 1797, Trevithick built a model steam carriage – and by 1801, he had built a full-size one, nicknamed the "Puffing Devil", which ran successfully in Camborne, Cornwall. The Puffing Devil was destroyed in an accident so, in 1802, Trevithick designed a locomotive that would run on rails. At the time, rails were used with horse-drawn wagons, mainly to transport coal from mines to ports for onward shipping. Trevithick's locomotive, built by the celebrated Coalbrookdale Ironworks, was possibly the first locomotive to run on rails.

## THE RAINHILL TRIALS

Although Richard Trevithick laid the foundations of the railways, it was not until the 1820s that people began to see steam trains as a serious alternative to horse-drawn transport. The first public railway designed from the start to use steam power was opened between Stockton and Darlington, in northern England, in 1825. Two of the shareholders and engineers on that first railway were father and son George (1781–1848) and Robert Stephenson (1803–1859).

In 1829, the Stephensons entered into the Rainhill Trials, a competition to find a locomotive for the forthcoming Liverpool and Manchester Railway. Their entry was called *Rocket*, and its many innovations made it the blueprint for all future steam locomotives. *Rocket* was the only locomotive to complete the ten 5-kilometre (3-mile) round trips required in the competition. When empty of cargo and passengers, it ran at a maximum speed of 47 kilometres per hour (29 miles per hour).

**OPPOSITE:** What remains of George and Robert Stephenson's original *Rocket* steam train, which is now on display in the Science Museum, London.

**RIGHT:** Oil on canvas painting of Richard Trevithick by John Linnell. (1792-1882). The painting shows Trevithick seated before a window, pointing to a view of the mountains.

**LEFT:** Trevithick built the first "flue boiler", in which hot exhaust gases pass through tubes inside the water tank and out through the tall chimney. The high-pressure steam produced in these boilers made possible more compact engines.

**ABOVE:** Artist's impression of Trevithick's London Steam Carriage of 1803, which was the world's first reliable self-propelled passenger-carrying vehicle. It had a top speed of about 15 kilometres per hour (9 miles per hour) on the flat, and weighed about a tonne when fully laden.

However, little is known about the locomotive, and only a single letter and drawing relating to it survive.

In 1803, Trevithick built another road vehicle, which he demonstrated in London. It attracted a lot of attention, but it was more expensive, noisier and more inconvenient than horse-drawn carriages, and went no further. In the same year, one of Trevithick's boilers exploded in Greenwich, London. This event could have set back his work; instead Trevithick invented a safety device, a "fusible plug", that he publicized but did not patent, in order to promote high-pressure steam.

The world's first steam train – carriages pulled by a locomotive – was the result of a bet. The owner of the Pen-y-Darren ironworks in Merthyr Tydfil, Wales, bet the manager of a neighbouring ironworks that a steam locomotive could be used to pull carriages filled with iron from his premises to a canal 16 kilometres (9 miles) away. The carriages were normally pulled by horses, so the rails already existed. Trevithick built a locomotive, and in February 1804, it successfully pulled 10.2 tonnes (10 tons) of iron and about 70 people the full distance. Although the rails broke in several places under the weight, the

concept of steam trains was proven. A year later, Trevithick built a lighter locomotive for a colliery in Newcastle, but although it worked, it was not put into service.

In 1808, Trevithick built a circular track in Euston, London, to promote the idea of steam trains. This was the world's first fare-paying passenger railway. From July to September that year, Trevithick's locomotive, the *Catch-Me-Who-Can*, ran around its track carrying passengers who paid five shillings for the privilege (later reduced to two shillings). It pulled a single carriage at speeds of about 20 kilometres per hour (12 miles per hour).

Trevithick also built a steam-powered dredging machine; powered a barge using one of his engines; and in 1812, he built an engine to thresh corn. In addition he invented an early propeller for steamboats and a device for heating homes, and he worked as an engineer on a tunnel under the River Thames in London, as well as on various projects in the silver mines of South America. But it is his pioneering contributions to the birth of the railways for which Richard Trevithick will be remembered.

## TIMELINE

**1603**
English entrepreneur Huntingdon Beaumont (c.1560–1624) builds the first railway – with horse-drawn wagons on wooden rails – to take coal from his mines near Nottingham, England.

**1767**
The first cast-iron rails produced at Coalbrookdale, England.

**1804**
Trevithick's Pen-y-Darren Locomotive pulls carriages with iron and 70 passengers along a stretch of 16 kilometres (9 miles) of track in Merthyr Tydfil, Wales – the world's first steam train.

**1807**
The first public railway – with a horse-drawn carriage – opens between Swansea and the Mumbles, Wales.

**1825**
The first public steam railway opens: it runs 40 kilometres (26 miles) between Stockton and Darlington, England.

**1829**
The Rainhill Trials, held at Rainhill, Merseyside, England, is a competition to find a locomotive for the nearly completed Liverpool and Manchester Railway.

**1872**
American inventor George Westinghouse (1846–1914) introduces the air brake. This uses a compressed-air system to apply brakes to wheels on all the carriages at the same time.

**1879**
German inventor Werner von Siemens (1816–1892) builds the first electric passenger train, in Berlin, Germany.

**1920s**
Diesel and diesel-electric locomotives begin to take over from steam locomotives.

**2004**
The first commercial maglev (magnetic levitation) railway – the Shanghai Transrapid Line – opens in Shanghai, China.

---

**OPPOSITE:** Trevithick's Coalbrookdale Locomotive – the world's first locomotive to run on rails. The Coalbrookdale was built for a colliery in Newcastle, in 1803. This contemporary illustration is the only source of information about it.

**TOP:** The Pen-y-Darren Locomotive, built in 1804, pulling wagons. The locomotive was formed by lifting one of Trevithick's existing stationary engines onto wheels at the Pen-y-Darren ironworks in Wales. It ran only three times, because it was too heavy for the iron rails. After the engine's trials, the railway returned to using horse power.

**ABOVE:** Trevithick's demonstration of the potential of steam trains in Euston, London, in 1808 – later called "The Steam Circus". The locomotive was called *Catch-Me-Who-Can*, because – to show that travel by steam would be faster – Trevithick raced it in a 24-hour race against horses, and won.

# Electromagnetism

## ELECTRIC MOTORS, GENERATORS AND TRANSFORMERS HAVE HELPED TO DEFINE THE MODERN WORLD.

English chemist and physicist Michael Faraday (1791–1867) made the first examples of each of these devices. More pure scientist than inventor, Faraday nevertheless had a practical bent, which led him to find innovative ways of using some of the incredible things he created in his laboratory.

Michael Faraday was born in Newington Butts, in London. Unlike most scientists of his day, he was not born into a wealthy family and did not benefit from much formal education. At the age of 13, his family secured an apprenticeship for him as a bookbinder.

Faraday took the opportunity to read many of the books he bound, and from these he developed an interest in science. In 1812, he was given tickets to a lecture by English chemist Humphrey Davy (1778–1829), who was about to retire

from the Royal Institution in London. Keen to move out of bookbinding, Faraday wrote up his notes from the lecture, bound them and presented them to Davy in the hope of being offered a job. When a position became available, Davy employed Faraday as his assistant.

**ABOVE:** Michael Faraday's Giant Electromagnet (also pictured top right). Faraday used this large magnet in his series of experiments testing electromagnetic fields.

**LEFT:** Faraday's Giant Electromagnet (1830), under the table in a mock-up of his laboratory at the Royal Institution, London. It was with this magnet that Faraday discovered that materials such as water and wood are repelled weakly by a strong magnet – a property called "diamagnetism".

**BELOW:** Replica of Faraday's induction ring – the world's first transformer, consisting of two long wires coiled around an iron ring. A changing electric current flowing in one coil produces a changing magnetic field in the iron ring, which induces a voltage in the other coil.

**BOTTOM:** Replica of apparatus used by Faraday in 1831 that changes movement energy into electrical energy. The permanent bar magnet moving in and out of the coil of wire "induces" a voltage in the solenoid. If the coil is part of a closed circuit, a current flows.

After Davy's retirement, Faraday travelled across Europe with him, meeting some of the most important scientists of the day. On his return, Faraday experimented in the field of chemistry, making several discoveries and inventing the earliest version of the Bunsen burner. A chance discovery in 1819/20 by Danish experimenter Hans Christian Ørsted (1777–1851) was to take Faraday in a new direction. Ørsted had discovered that whenever electric current flows, it produces magnetic forces. In 1821, Davy and his colleague William Wollaston (1766–1828) tried to use this phenomenon to make an electric motor, but they could not get it to work.

Later in 1821, Faraday succeeded where Davy and Wollaston had failed. He suspended a wire over a magnet in a cup of mercury. The wire rotated around the magnet whenever electric current flowed through it, because of the interaction between the magnetic field produced by the wire and the magnetic field of the magnet. Crude though it was, this was the precursor of all electric motors, which today are found in washing machines, drills and a host of other machines and appliances. When Faraday published his results, he failed to credit Davy, and the resulting fuss caused Faraday to stop working on electromagnetism until after Davy's death in 1829.

In 1831 in the basement of the Royal Institution, Faraday made a series of groundbreaking discoveries with batteries and wires. First, he discovered that a magnetic field produced by electric current in one wire can create, or "induce", electric current in another wire nearby. Faraday wound two long insulated wires around a circular iron ring, which intensified the effect; what he had made was the world's first transformer. Today, transformers are a vital part of the electricity distribution network, and they are also found in many home appliances, including mobile-phone chargers and televisions.

A month later, Faraday fixed a copper disc between the poles of a strong magnet and attached wires to the disc, one via the axle and one via a sliding contact. When he rotated the disc, an electric current was produced in the wires. This was the world's first electric generator. A year later, French instrument maker Hippolyte Pixii (1808–1835) read about Faraday's discovery and made an improved generator using coils of wire spinning close to a magnet's poles. Today, generators that supply huge amounts of electric power from power stations and wind turbines can trace their lineage directly back to Pixii's design.

In addition to his research and his inventions, Faraday instigated regular Friday discourses and the celebrated Christmas lectures at the Royal Institution; he himself was an inspiring lecturer. Later in his career, Faraday campaigned to clean up air and river pollution, and he was called upon to improve lighthouse technology and to investigate mining disasters. The most important contributions Faraday made, however, were those he made in the basement of the Royal Institution.

**ABOVE:** Michael Faraday lecturing at the Royal Institution. In 1825, Faraday instigated two series of public lectures that are still a feature of the institution: a series of Friday evening discourses and the annual Christmas Lectures, aimed at young people.

## JAMES CLERK MAXWELL
## (1831–1879)

During his researches with magnetism and electromagnetism, Michael Faraday became the first to describe "fields" of force. Several of his contemporaries expressed his discoveries in the precise language of mathematics, which Faraday's lack of formal education prevented him from doing. Most notable among these mathematical physicists was Scottish mathematician James Clerk Maxwell.

In the 1850s, Maxwell derived four equations that comprehensively describe the behaviour and interaction of electricity and magnetism. In 1864, Maxwell combined the equations, and the result was a single equation that describes wave motion. The speed of the wave described by the equation worked out to be exactly what experimenters had found the speed of light to be. Maxwell had shown that light is an electromagnetic wave. He went on to predict that light is a small part of a whole spectrum of electromagnetic radiation, a prediction that was confirmed in 1887 by the discovery of radio waves by German physicist Heinrich Hertz (1857–1894).

## TIMELINE

**1819** Hans Christian Orsted discovers that electric current produces magnetism.

**1820** French physicist André-Marie Ampère (1775–1836) discovers that current-carrying wires attract or repel each other.

**1821** Ampère invents the solenoid – a coil of wire that behaves just like a bar magnet when current flows through it.

**1821** Faraday makes the first electric motor.

**1824** English physicist William Sturgeon (1783–1850) invents the electromagnet.

**1831** American physicist Joseph Henry (1797–1878) makes an improved electromagnet, with tightly coiled, insulated wires, which can lift 340 kilograms (750 pounds).

**1831** Faraday makes several key discoveries about electromagnetism - including electromagnetic induction – and builds the first electromagnetic generator.

**1832** Hippolyte Pixii (1808–1835) makes an improved generator.

**1832** Baltic German diplomat Baron Pavel Schilling invents the first electromagnetic telegraph.

**1837** American inventors Samuel Morse (1791–1872) and his assistant Alfred Vail (1807–1859) invent the first practical long-distance electromagnetic telegraph system – the Morse telegraph.

**1864** James Clerk Maxwell publishes his electromagnetic theory of light, which shows that light is an electromagnetic wave.

63

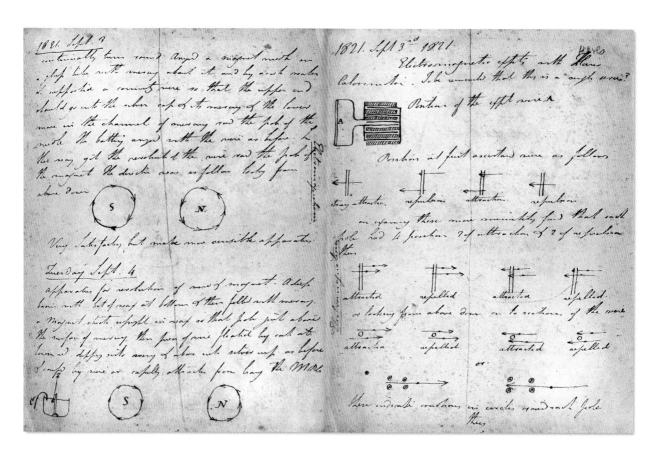

64

**ABOVE:** Pages from Faraday's laboratory notebook, September 1821, describing the world's first electric motor. Faraday describes how he suspended a wire in a basin of mercury. The wire rotated continuously around a magnet in the mercury whenever current flowed through it.

**RIGHT:** Photograph of Michael Faraday, taken by British photographer Henry Dixon before he set up his company Henry Dixon & Son Ltd in 1887.

1821. Sept 3

Here the wire moves in opposite circles and each pole or the poles move in opposite circles round the wire. I established the motion of the wire accurately being placed upright in a cork on water. Its lower end dipped into a table basin of mercury in the water and its upper enter into a little cavity silver cup containing a globule of mercury. the arrangement of poles always as at first. Magnets of different power held perpendicular to the wire did not make it revolve as Dr Wollaston expected but thrust it from side to side

The wire then bent into a crank form thus and by repeated applications of the poles of the magnets the following motions were ascertained. Looking from above down on the circle described by the bent part of the wire different Magnetic poles shewn by letters. North pole in centre.

The rod in the circle is merely just there to shew the point of the bent part

1821. Sept 3.

Magnetic poles on the outside of the circle the wire described.

The effect of the wire is always to pass off at a right angle from the pole indeed to go in a circle round it as when either pole was brought up to the wire perpendicular to it or to the radius of the circle it described there was neither attraction nor repulsion but the moment the wire moved in the above stated manner either in or out the wire moved one way or the other

The poles of the magnet act on the bent wire in all positions and not in the direction only of any axis of the magnet so that the current cannot can hardly be cylindrical or around the axis of any kind?

From the motion above a single magnetic pole in the centre of one of the circles should make the wire

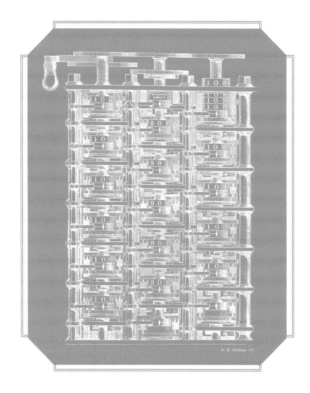

# The Mechanical Computer

LONG BEFORE THE INVENTION OF THE MODERN COMPUTER, A
DETERMINED GENIUS DESIGNED MACHINES THAT WOULD CARRY
OUT COMPLICATED MATHEMATICAL OPERATIONS, AND INVENTED
THE WORLD'S FIRST PROGRAMMABLE COMPUTING DEVICE.

The innovative mind behind the device was that of Charles Babbage (1791–1871), a brilliant mathematician, who also contributed to the development of business efficiency and railway travel.

As a child, Babbage was extremely inquisitive. In his autobiography, he wrote that whenever he had a new toy, he would ask his mother "What's inside it?", and would always break things open to find out how they worked. This curiosity gave him an early understanding of machines and mechanisms.

In 1810, he went to study mathematics at Trinity College, Cambridge University. At the time, mathematicians and engineers relied on books filled with tables of numbers to carry out calculations. There were tables of trigonometric functions

(sine, cosine and tangent) and tables of logarithms. The books contained hundreds of tables, and each table contained thousands of numbers. The values in the tables were worked

---

**ABOVE:** Babbage's Difference Engine No.1. It was built in 1832 by Joseph Clement, a skilled toolmaker and draughtsman. It was a decimal digital machine; the value of a number represented by the positions of toothed wheels marked with decimal numbers.

out by hand, by "computers" – a word that meant "people who compute". In 1812, Babbage moved college, to Peterhouse. In the library there, he realized that there were large numbers of mistakes in the numerical tables, and that they were down to human error. At the time, various mechanical calculating machines existed, but they were limited in what they could do. So Babbage envisaged a machine that would be able to calculate these tables at speed and remove the risk of human error.

In 1822, Babbage presented to the Royal Astronomical Society a proposal to build a calculating machine. The society granted Babbage money to set about making his machine, and he hired an engineer to oversee the job. In a workshop close to Babbage's house, with machine tools painstakingly designed by Babbage himself, the engineer set to work. It was an enormous task, and Babbage repeatedly asked for, and was granted, more money from the British Government.

Babbage called his proposed device the Difference Engine. It was never finished, because of a dispute between Babbage and the engineer – and perhaps also because it was so complicated. The Government officially abandoned the project in 1842. Babbage later improved his design, which he called Difference Engine 2. In 1991, London's Science Museum followed Babbage's design and constructed it; in 2005, they added a printer that had also been part of Babbage's original design. Both machines worked perfectly.

In 1827, his father, his wife and one of his sons died, and Babbage stopped work and took time to travel in Europe. While he was travelling, he dreamed up a more general calculating machine, which would be able to follow sets of instructions. Babbage envisaged a machine that would have input via punched cards, would be able to store answers, and would have a printer that would output the results. By 1835, he had produced the first of many designs for an "Analytical Engine" – the forerunner to the modern programmable computer. His design was expressed in 500 large engineering drawings, a thousand pages of engineering calculations and thousands of pages of sketches. Unfortunately, this machine was also never finished. In 1832, he published a book called *On the Economy of Machinery and Manufacture*, which was the

67

**ABOVE:** Charles Babbage was a notoriously difficult man, one of the many reasons given for the lack of realization of his designs.

**LEFT:** Babbage's collection of mathematical tables. His engines were designed to make these redundant.

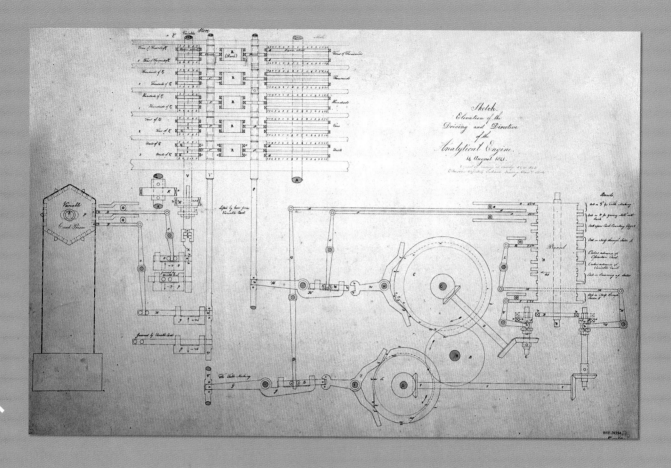

beginning of studies into the efficiency of business and industry. He also invented a special device to move objects off the railway track ahead of a train – affectionately called a cowcatcher.

Despite the fact that Babbage's Analytical Engine was never built in his lifetime, it created great interest. One of those who was inspired was mathematician Ada Lovelace (1815–1852), generally credited with being the world's first computer programmer. Lovelace met Babbage through her tutor, a mutual friend, and in 1842, she translated into English an Italian scientific paper on the Analytical Engine. She added copious notes to her version of the paper, and in those notes she described in detail a program that could run on the engine – and suggested that the engine could be programmed to do much more than simply calculate numbers.

**LEFT:** Babbage's cowcatcher in use on a steam locomotive in Pakistan's North-West Frontier Province. The concept was used on trains around the world.

**ABOVE:** A design sketched by Babbage for part of his Analytical Engine.

## ANALYTICAL ENGINE

Babbage's Analytical Engine was the first known design for a mechanical, general all-purpose computer. Although never built, the concepts it utilized in its design were at least 100 years before their time. Programs and data would be input using punched cards. Output consisted of a printer, a curved plotter and a bell. The machine's memory would be capable of holding 1,000 numbers of 50 decimal digits each. The programming language it was to use was very similar to that used in the early computers 100 years later. It used loops and conditional branching and was thus Turing-complete long before Alan Turing's (1912–1954) concept (see page 140). Although Babbage's direct influence on the later development of computing is argued greatly, Howard H. Aiken – the primary engineer behind IBM's 1944 Harvard Mark I (the first large-scale automatic digital computer in the United States) – said of Babbage's writings on the Analytical Engine, "There's my education on computers, right there: this is the whole thing, everything took out of a book."

## TIMELINE

**c. 2400 BCE**
The Roman abacus, one of the earliest examples of a mechanical counting machine, is used in Babylonia.

**c. 150–100 BCE**
The first analog computers are developed to perform astronomical calculations including the astrolabe in Ancient Greece.

**1206**
Al-Jazarī invents the first programmable computer – the castle clock. The length of day and night could be reprogrammed to compensate for the changing lengths of day and night throughout the year.

**1623**
German Wilhelm Schickard builds the first digital mechanical calculator.

**1643**
French philosopher Blaise Pascal invents the calculation device known as the Pascaline which is to be used in France to calculate taxes until 1799.

**1801**
Joseph Marie Jacquard introduces the concept of using punched cards to programme machinery, in this instance a weaving loom.

**1822**
Charles Babbage designs his Difference Engine – capable of holding and manipulating seven numbers of 31 decimal digits.

**1837**
Charles Babbage designs the first ever fully programmable mechanical computer: the Analytical Engine.

**1853**
Per Georg Scheutz builds a working Difference Engine using Babbage's design. It can create whole logarithmic tables mechanically.

**1880s**
Herman Hollerith creates a machine capable of large-scale automated data processing. The company later becomes IBM.

**1936**
Alan Turing gains his reputation as the father of modern computer science following his invention of the Turing Machine.

69

**RIGHT:** Design drawing, 1840, of Babbage's Analytical Engine, showing the incredibly complex arrangement of gears. Had it been built, this would have been the first truly automatic calculating machine. Babbage intended the machine to be powered by a steam engine.

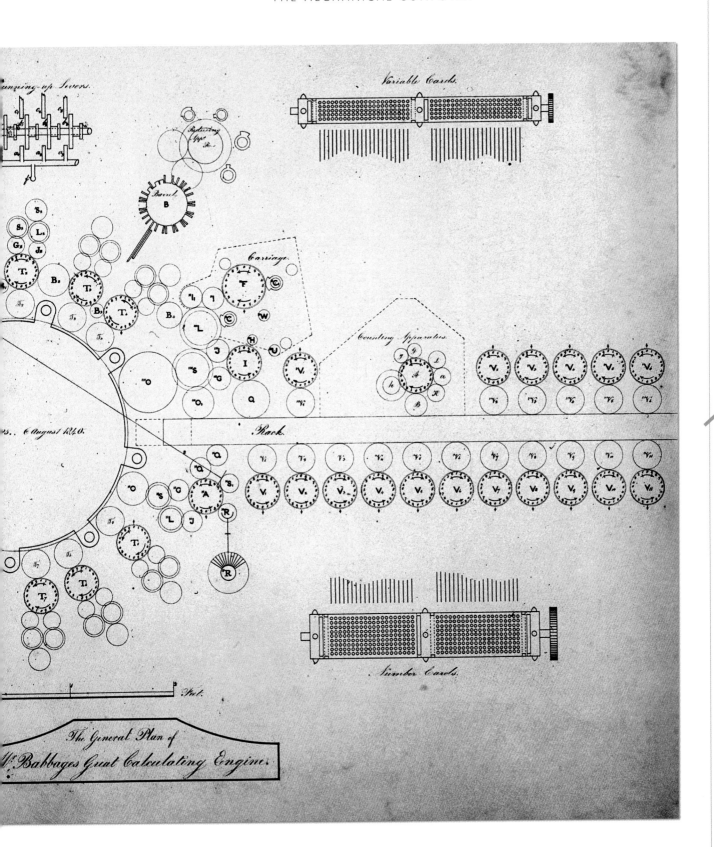

*Variable Cards.*

*Number Cards.*

*The General Plan of*
*Mr Babbages Great Calculating Engine.*

# Antiseptic

## UNTIL THE LATE NINETEENTH CENTURY, PATIENTS UNDERGOING EVEN MINOR SURGERY HAD ABOUT AS MUCH CHANCE OF DYING AFTERWARDS AS THEY DID OF SURVIVING.

English surgeon Joseph Lister (1827–1912) dramatically improved patients' chances in the 1870s, by introducing antiseptics into surgery.

Joseph Lister was born in Upton, in Essex, England, to a wealthy Quaker family. His father was a man of science, who made significant improvements to microscope design. Joseph studied the arts and then medicine at University College, London. Although born and educated in England, he spent most of his career in Scotland.

In 1856, Lister became an assistant surgeon at the Edinburgh Royal Infirmary. Four years later, he was appointed Professor of Surgery at Glasgow University Medical School. In 1861, Lister was put in charge of a new building with surgical wards at Glasgow Royal Infirmary. At the time,

around half of the patients died as a result of surgery – open wounds often festered, becoming badly infected and inflamed and full of pus. Untreated, this "wound sepsis" was often

**ABOVE:** French chemist and microbiologist Louis Pasteur in his laboratory, in a classic 1885 painting by Finnish-Swedish painter Albert Eledfelt (1854–1905). During the 1870s, Pasteur carried out a series of brilliant experiments that higlighted the existence – and effects – of airborne microbes.

life-threatening. The prevailing explanation of infection was
the so-called "miasma theory": the idea that polluted air was
the cause of disease. In the filthy air of the disease-ridden
cities of the nineteenth century, this was an easy connection
to make. But it badly missed the point; believing that polluted
air caused disease, surgeons carried out operations without
washing their hands and surgical wards were not clean.

In 1865, Lister read a report by French chemist and
microbiologist Louis Pasteur (1822–1895) suggesting that
fermentation and rotting are caused by airborne micro-
organisms. Pasteur also showed how micro-organisms can be
killed by heat, filtration or chemical attack. When Lister heard
of Pasteur's work, he realized that airborne micro-organisms
might be causing wounds to turn septic. He had heard that
carbolic acid (phenol, $C_6H_5OH$) had been used to stop sewage
from smelling bad, and had also been sprayed onto fields,
where it reduced the incidence of disease in cows. And so,
he and his surgeons began applying carbolic acid solution
to wounds, and using dressings that had been soaked in a
the same solution. In 1869, he developed a spray that would
fill the air with carbolic acid – the aim being to kill airborne

73

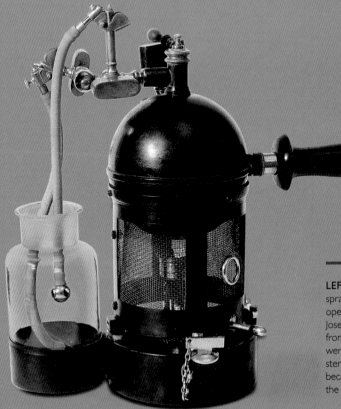

**LEFT:** Carbolic acid solution
spray, used to sterilize tools and
open wounds, as pioneered by
Joseph Lister. This example is
from France; French surgeons
were quick to adopt Lister's
sterile surgical procedures, in part
because it had saved many lives in
the Franco-Prussian war.

**ABOVE:** Glasgow slum, 1868.
As in all large cities at the time,
poor sanitation and overcrowding
led to the spread of infectious
diseases. This gave rise to the
miasma theory, in which "foul
air" was blamed for disease. The
miasma theory was eventually
superseded by the germ theory
of disease.

germs. Lister also told his surgeons to wash their hands before and after operations and to wash their surgical instruments in carbolic acid solution. His results were impressive: his surgical wards remained free of sepsis for nine months, and Lister had proved that carbolic acid was an effective antiseptic.

Other surgeons were slow to copy Lister's procedures, largely because many were reluctant to accept the idea that disease can be caused by micro-organisms – an idea known as the "germ theory of disease". When, gradually, surgeons did begin using his techniques, post-operative survival rates increased dramatically. It was after surgeons in the Franco-Prussian War of 1870–1871 used Lister's techniques, saving the lives of many wounded soldiers, that Lister's fame spread across Europe, and he began to receive the recognition he deserved. In 1877, Lister moved back to King's College, London, where he managed to convince many of the still-sceptical surgeons by successfully performing a complex knee replacement operation that had nearly always proved fatal. He continued to experiment tirelessly on improving surgical techniques and reducing mortality until his retirement in 1893.

Although Lister is famous for his antiseptic methods, he also worked on "aseptic" ones: attempting to keep operating theatres free of germs rather than killing them. Scottish surgeon Lawson Tait (1845–1899) defined modern aseptic surgical practices – even though he was not convinced of the existence of germs. Nevertheless, Lister's pioneering investigations into wound sepsis, his application of the germ theory of disease and his success in reducing mortality make his contributions to surgery of utmost importance.

**OPPOSITE:** Portrait of Ignaz Semmelweis on a 1965 Austrian stamp commemorating the hundredth anniversary of his death.

**BELOW:** Joseph Lister, centre, directing the use of his carbolic spray during a surgical operation, around 1865. Note the use of a cloth soaked in ether as an anaesthetic (left).

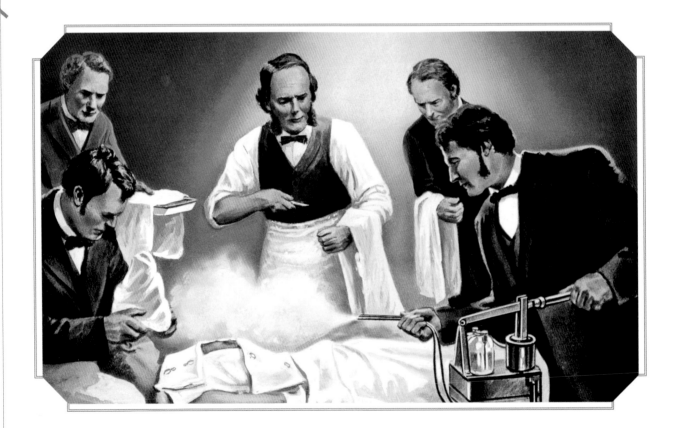

## IGNAZ SEMMELWEIS (1818–1865)

Nearly 30 years before Joseph Lister's pioneering work on antiseptic surgery, a Hungarian obstetrician, Ignaz Semmelweis, demonstrated the importance of washing hands. He worked in maternity wards at the Vienna General Hospital, in Austria. In wards attended by doctors and medical students, a disease called puerperal fever typically claimed the lives of about 20 per cent of women after childbirth, while in midwife-only wards, the incidence of puerperal fever was much lower.

Semmelweis realized that the doctors and students – who did not wash their hands between operations or even after dissecting corpses – were unwittingly transferring infections from one patient to another. In 1847, Semmelweis began a regime of washing hands with a solution of chlorinated water, and managed to reduce the mortality to below one per cent. Unfortunately, the medical community dismissed Semmelweis's results, and his work was quickly forgotten.

## TIMELINE

**1350s** Andalusian scholar Ibn Khatima (d.1369) attributes the infectious nature of the plague to "minute bodies".

**1546** Venetian physician Girolamo Fracastoro (1478–1553) suggests that infectious diseases are caused by "disease seeds" - tiny particles carried in the air or from contact with infectious objects.

**1670s** Antonie van Leeuwenhoek (1632–1723) becomes the first person to observe micro-organisms (in pond water) using his powerful, single-lens microscopes.

**1826** Lister's father, English physicist Joseph Jackson Lister (1786–1869), greatly improves the microscope, which leads to an explosion of interest in and research into micro-organisms.

**1847** Ignaz Semmelweis reduces the incidence of puerperal fever in his maternity wards by instigating a regime of inter-operative cleanliness.

**1854** English physician John Snow (1813–1858) disproves the miasma theory, for cholera at least, by showing that the disease is transmitted by something in water, not by foul air.

**1860s** Louis Pasteur carries out a series of investigations that prove "germs" are the cause of food going bad; he also proves that a disease in silkworms is caused by micro-organisms.

**1865** Joseph Lister pioneers sterile surgery, introducing carbolic spray and encouraging surgeons to wash their hands.

**1880s** Lawson Tait establishes modern sterile surgical practices, by encouraging asepsis (absence of germs) in operating theatres, rather than antisepsis pioneered by Lister.

**1890** German physician Robert Koch (1843–1910) puts the germ theory beyond doubt and establishes criteria that enable investigators to discover which diseases are caused by micro-organisms and which are not.

# The Motor Car

THE PERSON RESPONSIBLE FOR DESIGNING THE FIRST TRUE
MOTOR CAR, GERMAN ENGINEER KARL BENZ, HAD NO IDEA
WHAT EFFECT HIS INVENTION WOULD HAVE ON THE WORLD.

By increasing mobility, less than 100 years after the rise of the railways, the motor car once again revolutionized patterns of work and play and the distribution of goods, and its rapid uptake in the twentieth century changed the landscape quickly and dramatically.

Karl Benz (1844–1929) was born in Karlsruhe, Baden (now in Germany). His father died when Karl was just two years old, but his mother encouraged him greatly, working hard to put him through grammar school and the Karlsruhe Polytechnische Schule (Institute of Technology). It was his dream from early on to invent a form of transport that would run without horses and off rails. The idea of self-propelled road vehicles was already popular before Benz was born. Some engineers had made "cars" – mostly steam carriages and electric vehicles; all of them were adaptations of horse-drawn carts and none was particularly

effective. The most crucial invention in the development of the motor car was the internal combustion engine. In a steam engine, the combustion – the fire that heats the steam – is produced outside the cylinder. The first practical engines in which

**ABOVE:** Photograph of the original, and unique, Benz Patent Motorwagen, 1886. The car was converted to a four-wheel vehicle in the 1890s, then in 1903, it was returned to its original form. It is now on display at the Deutsches Museum, in Munich, Germany.

**OPPOSITE:** Replica of Benz's patent motor car, showing the single-cylinder, four-stroke engine, horizontal flywheel and belt drive. The original ran on ligroin, a mixture of hydrocarbons very similar to petrol. Also visible are the fuel tank, in the foreground, and the cooling water tank.

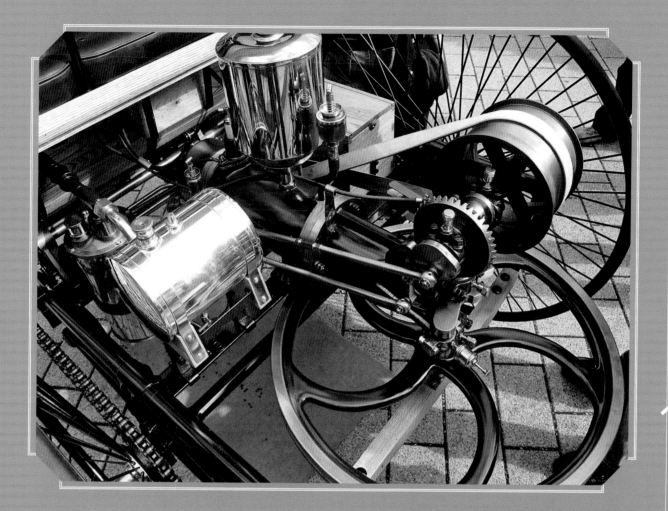

### HENRY FORD (1863–1947)

For 20 years after Karl Benz's Patent Motorwagen, motor cars were not available to most people. The fact that each one had to be made individually kept the cost high, which in turn kept demand low. In 1908, American entrepreneur Henry Ford set out to change that, when he introduced what he called "a car for the great multitude".

The affordable Ford Model T was a real breakthrough, being made from interchangeable parts in a factory with tools laid out in an efficient arrangement. From 1913, the cars were manufactured on assembly lines. One of Ford's employees had seen how effective production lines could be when he visited a meat-packing factory in Chicago. The application of the idea to the motor-car industry brought costs down dramatically, made Henry Ford incredibly rich and had a rapid and profound effect on the world of the twentieth century.

combustion took place inside the cylinder, and drove a piston directly, appeared in the 1850s. The most important was invented in 1859 by Belgian engineer Étienne Lenoir (1822–1900).

The next step towards motor cars proper was the "four-stroke" engine designed by German inventor Nikolaus Otto (1832–1891) in 1876. The four strokes – intake of the fuel-air mixture; compression of that mixture; ignition; and exhaust – still form the basis of petrol engines today. Otto's engine was the first real alternative to the steam engine.

Karl Benz closely followed developments in engine design after leaving college, and worked towards his dream of building a motor car. He had been employed on various mechanical engineering projects, and in 1871 had moved to the nearby city of Mannheim. In the 1870s, Benz designed a reliable two-stroke petrol engine (in which the four operations of the four-stroke engine are combined into one upward and one downward stroke), for which he was granted a patent in 1879. Four years later, he formed a company with two other people: Benz & Company Rheinische Gasmotoren-Fabrik. The company began as a bicycle repair shop, and quickly grew when it began making machines and engines.

Benz & Company did well, giving Benz the time and the confidence he needed to pursue his dream. By the end of 1885,

Benz's car was ready. It was a three-wheeled carriage powered by a single-cylinder four-stroke engine, which he had created specially. Benz's motor car incorporated several important innovations of his own design. These included an electric starter coil, differential gears, a basic clutch and a water-cooling system for the engine. Despite his hard work and his obvious brilliance, Benz had not quite worked out how to achieve steering with four wheels. He took the easy option and had three wheels, the single front wheel being turned by a "tiller"-type handle.

Benz applied for a patent in January 1886, and it was granted in November of that year. His application was successful because his motor car had been designed from the start as a self-powered vehicle, and not simply as a cart with an engine attached.

After a few improvements, including the world's first carburettor, the first Benz Patent Motorwagen was sold in 1887. Benz began production of the car, and advertised it for sale in 1888; it was the first commercially available production car in history. Uptake was very slow, however, so Benz's wife Bertha (1849–1944) decided to try to raise awareness. In August 1888, she drove with her two sons from Mannheim to her home town of Pforzheim and back – a total distance of nearly 200 kilometres (120 miles). of Pforzheim and back. The journey was not trouble free. Benz had to improvise new electrical insulation with

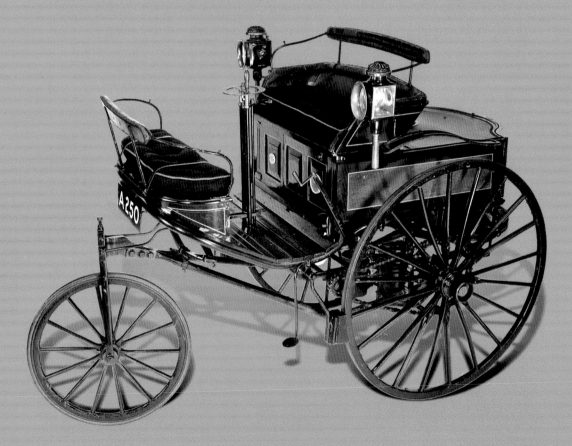

her garter, clear a blocked fuel pipe with a hat pin and, when the vehicle's wooden brakes jammed, get a cobbler to install protective leather strips over them. The world's first brake pads had been invented. The stunt generated a great deal of publicity – and thanks at least in part to that publicity, Benz's Motorwagen became a real success. The age of motoring had begun.

The men who worked so industriously on early versions of the motor car imagined sunny days of summer driving, hazard-free from weather or other motorists. Non-driver Mary Anderson (1866–1953) was able to see past the hype of halcyon travel, imagine the dangers presented by the real world and offer a solution we still use today.

Born in Alabama, Anderson was always looking for new projects. She tried real estate development, cattle ranching and even running a vineyard but when she returned to Birmingham in 1893 to care for an elderly aunt she had a brainwave. The brainwave came when Anderson was visiting friends in New York City and noticed that, when it rained or snowed, trolley-car drivers had to either get out and clean their windscreens or open the windows and stick their heads outside. The best remedy anyone had come up with so far was a "split windshield" that would partially open, but still blasted the driver with freezing, wet wind.

Anderson went home and started drawing a design that even today we would recognise as a windscreen wiper. Operated by the driver turning a lever inside the car, rubber blades would wipe across the screen, clearing as they went.

She commissioned a local company to mock up a prototype and, in 1903, patented her car-window cleaning device for 17 years. The product was met with scepticism, if not outright derision – people thought that in the very least it would be dangerous distraction for a driver. She applied to a Canadian firm to see if they would install her wipers, but they claimed it was of no commercial value.

Disappointed, Anderson let the patent drop in 1920, just as windshield wipers, of extremely similar design to hers, were becoming a standard part of automobile manufacture.

By this point though, technology was going automatic. Canadian stage actress Charlotte Bridgwood (1861–1929) had come up with the Storm Windshield Cleaner in 1917, which used rollers instead of blades and was powered by electric motor. Her daughter, famous silent movie actress and motor car-fanatic Florence Lawrence (1886–1938), may have been involved in the design of the Storm Windshield Cleaner, but was also an inventor in her own right, coming up with the first mechanical turning and braking signals for cars. She never patented the invention and neither she nor her mother got much more credit than Mary Anderson. Margaret A. Wilcox (b.1838) was more fortunate. In 1893 she pioneered the first car heater.

## TIMELINE

| | |
|---|---|
| 1859 | Étienne Lenoir invents the first practical internal combustion engine. |
| 1876 | Nikolaus Otto invents the four-stroke internal combustion engine. |
| 1886 | 1886 Karl Benz applies for a patent for the motor car. |
| 1892 | French-born German inventor Rudolf Diesel (1858–1913) develops the diesel engine. |
| 1901 | German engineer Ferdinand Porsche (1875–1951) invents the hybrid electric car, driven by a petrol engine and an electric motor. |
| 1902 | French industrialist Louis Renault (1877–1944) invents the drum brake, still found on most cars today. |
| 1902 | English inventor Frederick Lanchester (1868–1946) invents the disc brake, still almost universal on cars today. |
| 1903 | American inventor Mary Anderson (1866–1953) invents the modern swinging-arm rubber windscreen wiper. |
| 1908 | Henry Ford introduces the Ford Model T, the first mass-produced, affordable car. |
| 1910 | American inventor Charles Kettering (1876–1958) invents all-electric ignition. |
| 1913 | Henry Ford introduces the first motor-car production line. |
| 1924 | The world's first motorway, the *autostrada dei laghi*, opens in Italy. |

**OPPOSITE:** By 1888, Benz had improved his design, and began producing cars in greater numbers. French engineer and entrepreneur Émile Roger, in Paris, held the sole rights to sell Benz's cars outside Germany, and helped to popularize the vehicle.

**RIGHT:** German patent number DRP 37435 for Benz's gas-powered engine, which was intended for "light vehicles and small vessels to transport between one and four persons". The document includes a detailed drawing of Benz's main use for the engine: his motor car.

IMPERIAL PATENT OFFICE
PATENT NO. 37435
(ISSUED ON 2ND NOVEMBER, 1886)

CLASS 46: AIR-POWERED AND GAS-POWERED MACHINES

BENZ & CO. in MANNHEIM

Vehicle with Gas Engine Drive

Patented in the German Empire as from 29th January, 1886

The Discovery relates to the operation of mainly light carriages and small vessels, such as are used to transport 1 to 4 persons. The appended drawing shows a small tricycle-like vehicle, constructed for 4 persons. A small gas engine (any system can be used) serves as the motive power. The latter receives its gas from an accompanying apparatus, in which gas is generated from petroleum ether or from other gasifying materials. The engine's cylinder is kept at constant temperature through the evaporation of water.

The engine is laid out in such a way that its flywheel rotates in a horizontal plane and the power is transmitted through two bevel gears to the main wheels. This not only makes the vehicle fully manoeuverable, but also provides a safeguard against any tipping over of the same when being driven around small curves, or should there be any obstructions on the route.

80

KAISERLICHES PATENTAMT.

# PATENTSCHRIFT

## № 37435

KLASSE 46: LUFT- UND GASKRAFTMASCHINEN.

### BENZ & CO. IN MANNHEIM.

**Fahrzeug mit Gasmotorenbetrieb.**

Patentirt im Deutschen Reiche vom 29. Januar 1886 ab.

Vorliegende Construction bezweckt den Betrieb hauptsächlich leichter · Fuhrwerke und kleiner Schiffe, wie solche zur Beförderung von 1 bis 4 Personen verwendet werden.

Auf der beiliegenden Zeichnung ist ein kleiner Wagen nach Art der Tricycles, für 2 Personen erbaut, dargestellt. Ein kleiner Gasmotor, gleichviel welchen Systems, dient als Triebkraft. Derselbe erhält sein Gas aus einem mitzuführenden Apparat, in welchem Gas aus Ligroin oder anderen vergasenden Stoffen erzeugt wird. Der Cylinder des Motors wird durch Verdampfen von Wasser auf gleicher Temperatur gehalten.

Der Motor ist in der Weise angeordnet worden, dafs sein Schwungrad in einer horizontalen Ebene sich dreht und die Kraft durch zwei Kegelräder auf die Triebräder übertragen wird. Hierdurch erreicht man nicht nur vollständige Lenkbarkeit des Fahrzeuges, sondern auch Sicherheit gegen ein Umfallen desselben beim Fahren kleiner Curven oder bei Hindernissen auf den Fahrstrafsen.

Die Kühlung des Arbeitscylinders des Motors geschieht durch Wasser, welches die ringförmigen Zwischenräume ausfüllt. Gewöhnlich läfst man das Kühlwasser bei Gasmotoren mit geringer Geschwindigkeit durch den Cylinder sich bewegen, indem das kalte unten eintritt und das erwärmte oben abfliefst. Es ist aber dazu ein grofser Wasservorrath nöthig, wie ihn leichte Fuhrwerke zu Land nicht gut mitführen können, und daher folgende Einrichtung getroffen worden: Das Wasser um den Cylinder verdampft. Die Dämpfe streichen durch das oberhalb des Cylinders angebrachte Rohr-

system 1, werden dort zum gröfsten Theil condensirt und treten wieder als Wasser unten in den Cylinder ein. Der nicht condensirte Dampf entweicht durch die Oeffnung 2.

Das zum Betrieb des Motors nöthige Gas wird aus leicht verdunstenden Oelen, wie Ligroin, dargestellt. Um stets ein gleichmäfsiges Gasgemenge zu erhalten, ist es nöthig, dafs neben dem gleichmäfsigen Luftzutritt und der gleich hohen Temperatur des Ligroins auch der Stand des letzteren im Kupferkessel 4 ein möglichst gleicher sei, und ist zu diesem Zweck der Vorrathsbehälter 5 mit dem Kupferkessel 4 durch eine enge Röhre 6, die in ein weites Wasserstandsglas 7 mündet, verbunden. An der Röhre ist ein kleiner Hahn 8 angebracht, um den Zuflufs nach Bedarf reguliren zu können. Durch die Glasröhre ist das tropfenweise Eintreten des frischen Ligroins wahrzunehmen und zugleich der Stand desselben im Apparat zu controliren.

Das Ingangsetzen, Stillhalten und Bremsen des Fuhrwerkes geschieht durch den Hebel 9. Der Motor wird, bevor man den Wagen besteigt, in Betrieb gebracht. Dabei steht der Hebel f Mitte. Will man das Fuhrwerk in Bew ng setzen, so stellt man den Hebel 9 nach v wärts, wodurch der Treibriemen vom Leerla1 auf die feste Scheibe geschoben wird. Beim Anhalten bewegt man den Hebel 9 wieder auf Mitte, und will man bremsen, so drückt man ihn über Mitte rückwärts. Der ausgerückte Riemen bleibt dabei in seiner Stellung und nur die Bremse wird angezogen. Um zu bewirken, dafs, wenn der Riemen auf Leerlauf gestellt ist, derselbe bei weiterer Rück-

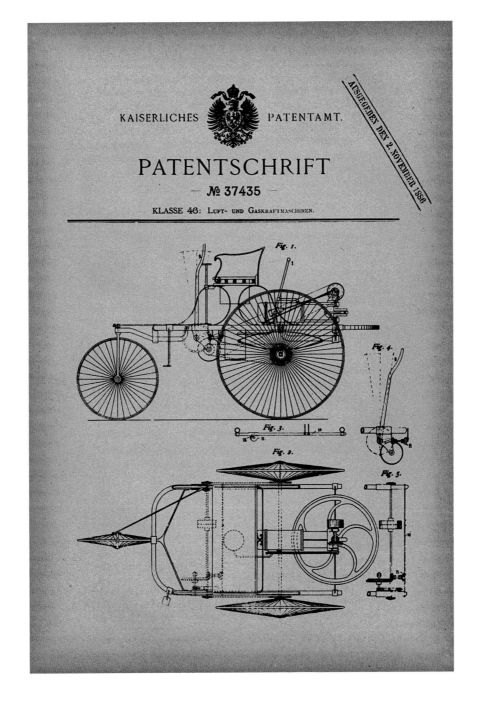

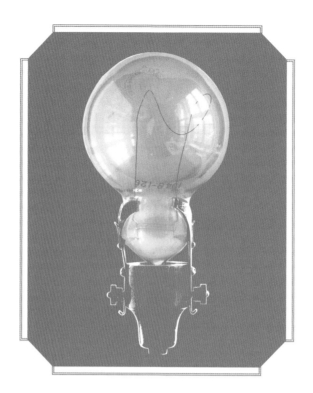

# The Light Bulb

IT'S THE UNIVERSAL SYMBOL OF THE "INSTANT IDEA", BUT IN
REALITY IT TOOK MANY MINDS AND NEARLY A CENTURY TO
COME UP WITH THE ORIGINAL "LIGHT BULB MOMENT".

Thomas Edison (1847–1931) was granted a total of 1,093 US patents, but perhaps his greatest invention of all was something he could not patent: organized, systematic research.

Home-schooled from the age of 12, Edison set up his first laboratory in his bedroom in Port Huron, Michigan. Much of his early effort was dedicated to improving the telegraph, a system that had revolutionized long-distance communication. At 14, Edison built a working telegraph at home; by 16, he was working as a telegraph operator at his local telegraph office.

Edison's first successful invention was the "Universal Stock Ticker" (1870) – a device with which dealers could receive the current share prices from the New York Stock Exchange. With the money he earned from selling the rights, he set up a workshop in Newark, New Jersey.

In 1873, he invented the "Quadruplex Telegraph", allowing the simultaneous transmission and reception of four telegraph signals on a single wire. Edison sold the rights to Western Union – saving them millions of dollars in wiring – and the proceeds helped him move to a new 34 acre (14 hectares) site in Menlo Park, New Jersey, where he set up a full research and

**ABOVE:** An "Ediswan" lamp, c.1890. English physicist Sir Joseph Wilson Swan (1828–1914) took Edison's lamp and improved upon it further.

development laboratory – the first of its kind anywhere in the world. At Menlo Park, Edison invented a sensitive microphone, filled with compressed carbon, which improved the distance over which telephone calls could be made. As an offshoot of his research into the telephone, Edison and his team invented a device for recording sound: the phonograph.

Perhaps the most important invention to come out of Menlo Park, however, was the light bulb. As is true of nearly all his inventions, Edison did not actually invent the light bulb; he made significant improvements that made it practicable for the first time. His use of a carbonized bamboo filament meant a bulb would light for 40 hours instead of just a few minutes. He demonstrated the new technology in December 1879, lighting the workshop in a public demonstration.

Edison set up a bulb-making factory at Menlo Park, and his success with electric light led him to work on a system to distribute electric power. He patented the system in 1880, and by 1882, he had set up a power station at Pearl Street, New

York. His research team went on to invent the first device for showing moving pictures (using 35mm sprocketed film, which became the film industry standard), a new type of battery, a device for separating iron ore, an all-concrete house, and an electric locomotive.

**ABOVE:** The galvanizing room in Edison's Menlo Park laboratory. His early electric bulbs can be seen on the table.

## INCANDESCENT LIGHT BULB

The first electric lighting depended upon a phenomenon known as incandescence. This is the production of light by objects because they are hot. The element of a toaster glows orange by incandescence, for example. In an incandescent light bulb, electric current passes through a thin piece of conducting material called a filament. The filament becomes extremely hot, and glows yellow- or white-hot. The problem to be overcome in the late nineteenth century was ensuring that the filament did not simply burn. Edison, and Joseph Swan, overcame this problem by evacuating the glass bulb (removing the air). In the early twentieth century, light bulbs were filled with inert (unreactive) gases instead.

84

The light bulb Edison invented was an incandescent lamp (see box, opposite). This technology dominated electric lighting for all of the twentieth century, as it is simple and cheap to make. However, incandescent bulbs are extremely inefficient: most of the energy supplied to them is lost as heat. Also, these bulbs are notoriously short-lived: the filament is fragile and does eventually burn through.

A rival, less wasteful, form of electric lighting, the fluorescent lamp, was developed in the 1930s. Concerns about energy efficiency from the 1990s onwards led designers to make compact fluorescent lamps (CFLs) as a direct replacement for incandescent bulbs. Nowadays, the incandescent bulb is rare, having been largely usurped by CFLs and the even more efficient light emitting diode (LED) lighting.

After Edison died, US President Herbert Hoover encouraged Americans to turn off their lights for one minute, in tribute to the contributions made by America's greatest inventor.

**OPPOSITE:** The Dynamo Room at Pearl Street Station, New York. Pearl Street, the first central power plant in the US, was built by Edison's Electric Illuminating Company and started generating electricity on 4 September 1882. By 1884, it was serving 508 customers and powering 10,164 lamps.

**ABOVE:** Photograph of Thomas Alva Edison, c.1880, shortly after the creation of his first practicable light bulb.

## TIMELINE

**1802** Sir Humphrey Davy (1778–1829) creates the first incandescent light by passing an electric current through a platinum filament. It is neither bright nor long-lasting enough to be practicable but forms the basis of research into electric lighting for the next 75 years.

**1809** Davy invents the first carbon arc lamp.

**1840** Warren de La Rue (1815–1889) places a coiled platinum filament in a vacuum tube enabling longer burn time. The use of platinum makes it commercially unviable.

**1841** Frederick de Moleyns is granted the first patent for an incandescent lamp – the design uses platinum wires encased in a vacuum bulb.

**1851** Jean Eugene Robert-Houdin (1805–1871) publicly demonstrates incandescent light bulbs on his estate in Blois, France.

**1874** Henry Woodward and Matthew Evans patent their lamp in Canada using carbon rods encased in a nitrogen-filled glass cylinder.

**1875** Joseph Wilson Swan builds on his use of the cheaper carbon fibre filament in the 1860s, instead of platinum. His house in Gateshead, north-east England, was the first in the world to have working light bulbs installed.

**1878** Swan obtains a patent on his carbon filament bulb, a year before Edison.

**1879** Thomas Edison invents the first commercially practicable incandescent lamp. His bulb can last for over 40 hours.

**1910** William David Coolidge (1873–1975) invents the first commercially viable tungsten filament, capable of outlasting all other types of filament. This tungsten-filament bulb forms the basis for all future luminescent bulbs.

85

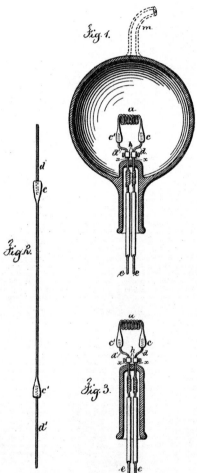

T. A. EDISON.
Electric-Lamp.

No. 223,898.          Patented Jan. 27, 1880.

Fig. 1.

Fig. 2.

Fig. 3.

Witnesses
Chas. H. Smith
Geo. T. Pinckney

Inventor
Thomas A. Edison
for Lemuel W. Serrell
Atty

## THE DIFFICULT BIRTH OF THE LIGHT BULB

Like Thomas Edison, Joseph Swan (1828–1914) did not invent the light bulb from scratch. Many other physicists beforehand had experimented with ideas, not least Ebenezer Kinnersley (1711–1778), a neighbour of Benjamin Franklin who, in 1761, heated a wire to incandescence, discovering "electric fire" and Sir Humphry Davy (1778–1829) who managed to create a dim light for a short time by passing a current through a strip of platinum. Dozens more inventors, inspired by others' success, continued the work.

Joseph Swan (1828–1914) developed a method of processing filament through a vacuum to lengthen the lifetime of a bulb. He used a thin carbon rod, which wasn't practical for commercial use, but after much experimentation he managed to create a filament that could last about 40 hours. He patented the device in 1880, the same year his house, at Low Fell, Gateshead became the first in the world lit by electricity. In 1881, the luxurious new Savoy Theatre in London became the first public building in the world to be lit entirely by electricity.

**OPPOSITE:** Diagram from Edison's 1880 patent for an "electric lamp". Edison's main innovation was to use a long, coiled filament; figure 2 shows the filament before coiling. Coiling allowed Edison to fit a long filament inside a small bulb, dramatically increasing the filament's resistance.

**ABOVE:** A modern lightbulb, which has come a long way since Edison's "lamps" but still very closely resembles them.

# The Telephone

PROBABLY THE MOST LUCRATIVE PATENT OF ALL TIME WAS AWARDED TO A SCOTTISH-CANADIAN-AMERICAN INVENTOR IN 1876, FOR A DEVICE THAT HAD THE MAGICAL ABILITY TO TRANSMIT THE SOUND OF THE HUMAN VOICE ACROSS LONG DISTANCES.

The inventor's name was Alexander Graham Bell (1847–1922), and the device was the telephone. Bell was born in Edinburgh, Scotland. His father and grandfather were pioneers in the field of speech and elocution, and his mother had a condition that resulted in progressive hearing loss. These influences inspired Bell to study language and the human voice. The young Bell attended a prestigious school in Edinburgh, and when he left aged 16, he taught music and elocution before studying in Edinburgh and London. After his studies, Bell taught deaf people to speak, using a method his father had developed, and it was during this time he began experiments in the transmission of sound using electricity.

**ABOVE:** The first wireless telephone call, using a photophone transmitter, in April 1880. Sound made a tiny mirror vibrate inside the transmitter, which changed the intensity of a light beam reflected off it. At the receiver, a light-sensitive cell detected the changes in intensity and reproduced the speech sounds.

Bell lost both his brothers to tuberculosis, and in 1870 his own precarious state of health deteriorated. His parents decided the family should emigrate to Canada. Within a year of arriving, Bell had become a Canadian citizen, and his health had improved. The family settled on a farm, and Bell continued his experiments with sound and electricity. He spent time teaching deaf people in Montreal, Canada, and in various American cities. Eventually, he settled in Boston, where he founded a school for the deaf and became professor of vocal physiology at Boston University. However, in 1873, becoming increasingly preoccupied with his attempts to transmit sound with electricity, he resigned his position. He retained two deaf people as private students; as luck would have it, their wealthy parents became interested in what he was trying to achieve, and helped fund his work.

By 1874, Bell had built a device called a harmonic telegraph, which was designed to transmit several telegraph messages at the same time through a single wire. Each message was sent as pulses of electricity with a distinct frequency of alternating current. Bell's financial backers were keen for him to perfect his device, but Bell was much more interested in trying to adapt his

89

**ABOVE:** Replica of Bell's 1875 experimental telephone transmitter. Speech sounds caused the stretched parchment drum to vibrate, and the metal spring with it. A magnet attached to the spring produced an alternating electric current in the coil above it – one that matched the vibrations of the sound waves.

**RIGHT:** In 1887, news of Bell's success in transmitting speech spread worldwide. Britain's new Queen Victoria asked Bell to demonstrate it at her residence on the Isle of Wight. This telephone and "terminal panel" were part of the resulting installation.

## ELISHA GRAY (1831–1901)

Alexander Graham Bell's company fought a total of 587 lawsuits over priority in the invention of the telephone during the 1880s and '90s. The company won them all, ultimately due to the fact that no one had claimed priority until many months after Bell was awarded his patent. However, some controversy remains over Bell and one of his competitors at the time: prolific American inventor Elisha Gray.

On the same day as Bell filed his patent, 14 February 1876, Gray filed a patent "caveat" at the same office, for a very similar device. There is evidence that Bell had sight of Gray's application. In Bell's first successful experiment, he used a water-based microphone Gray had designed. But he never used it in public demonstrations, probably because he knew he should not have known about it. Instead he used his own, less effective, electromagnetic receiver.

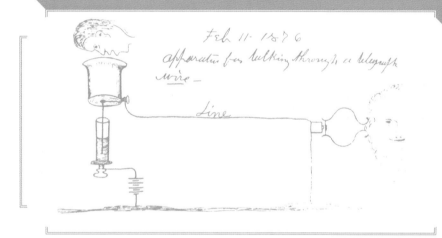

90

**OPPOSITE:** Alexander Graham Bell, seated, in New York, on 18 October 1892, at the opening of the first long-distance telephone service. The line connected New York and Chicago, in the USA: a distance of about 1,140 kilometres (710 miles).

device to transmit the human voice through a wire, something that many thought was impossible. In 1875, Bell was getting close, but his knowledge of electricity was lacking. Fortunately, that year he met an electrical technician called Thomas Watson (1854–1934), whom he engaged as his assistant.

When Bell was granted the patent for the telephone, his device had not yet transmitted any speech. But three days later, on 10 March 1876, Bell and Watson achieved success. Bell, in one room, spoke into the device, and in an adjoining room, Watson heard the now famous words, "Mr Watson, come here – I want to see you". In the following months, Bell and Watson made improvements to the microphone, and his device transmitted speech over increasing distances – and began to generate huge interest from scientists and the press. In 1877, he and his financial backers formed the Bell Telephone Company.

Bell's inventions were not restricted to the telegraph and the telephone. He improved Edison's most famous creation: an early sound-recording device called the phonograph. He also invented record-breaking speedboats that rose up out of the water on submerged "wings" called hydrofoils, a chamber to help people with respiratory problems breathe (an early version of the iron lung) and the first metal detector. In his later years, he spent a great deal of time and effort experimenting with flight. The invention of which he was most proud, however, was the photophone, a device that transmitted sound using light rather than electricity. In 1880, Bell's photophone made the first ever wireless transmission of speech, across a distance of more than 210 metres (230 yards). Although his idea never took off at the time, it is similar to the way telephone signals are transmitted today using laser light passing through optical fibres.

Jazz bass player Teri Pall (1921–2013) held more than 33 patents for the various inventions she devised during her creative lifetime. These included a solar-powered cooker, a wrist chronograph calculator and a miniature electronic pain blocker. The most significant of them was undoubtedly the cordless phone she prototyped in 1965.

Pall's cordless phone consisted of a base unit and a remote handset that worked on low radio frequencies; it had a range of two miles. Paradoxically, it was this range that made the device unmarketable, since the radio signals the phone generated interfered with aircraft communications. Pall sold the manufacturing rights to her invention in 1968.

The cordless phone was not the only invention Pall devised that involved radio. According to her own account, her very first discovery, which she made when aged only 12, involved her family radio set. As she recalled, the set was receiving two stations simultaneously, so she took it to bits and modified the tuning device to solve the problem. She showed her modification to the radio's manufacturer in New York; after only a few minutes' haggling, she was paid $2,000 for the modification.

## TIMELINE

**1861** The German inventor Johann Philipp Resi (1834–1874) demonstrates a near-working electromagnetic telephone; it was later shown that his apparatus could transmit speech, though very weakly.

**1876** Bell applies for a patent for the telephone – on the same day as Elisha Gray. Three days later, Bell speaks the first ever words transmitted successfully by telephone.

**1878** English inventor Henry Hummings (1858–1935) produces the first effective microphone for the telephone, filled with carbon granules.

**1878** The first telephone directory – a single page – is published in New Haven, Connecticut.

**1882** American inventor Leroy Firman invents the switchboard; until then people rented telephones in pairs which could only connect with each other.

**1889** American inventor Almon B. Strowger (1839–1902) invents an electromechanical 'stepping switch', bringing rotary dialling and automatic exchanges.

**1892** American Bell Telephone Company launches the first long-distance telephone service, between New York and Chicago.

**1941** American company AT&T launches the first touch-tone telephone, replacing rotary dialling.

**1960s** Digital telephony – in which sound is digitized for ease of routing – begins to take over from analogue services in most countries. Digital exchanges offer a wider range of services than analogue ones.

**1979** Japanese company NTT launches the first "cellular" mobile telephone service, in Japan.

**2000s** VoIP (Voice over Internet Protocol) services, routing calls as Internet traffic, begin to compete with the POTS (Plain Old Telephone Service).

46

## Saturday March 11th 1876

Fig I

1. Mr Watson completed the Receiving Instrument shown in Fig I this afternoon. C is a wooden capsule. SS' a steel spring armature. E Electro-magnet. NN' an iron a hollow iron cylinder within which the electro magnet E is placed. P P' the core. The pole P is positively Magnetized — the circular rim NN' negatively charged.

The instrument was tried this afternoon and no audible effect was heard at A.

2. The capsule C was removed and the ear applied directly to the spring SS', a clean sound was perceptible. These experiments were made with a tuning fork as shown in Fig. 2 page 43. The above instrument taking the place of the Receiver S (Fig 2 page 43)

M.G.H. March 12 1876          Noted by A.G.B. March 12th 1876

---

37

## March 9th 1876

Fig I

1. The apparatus suggested yesterday was made and tried this afternoon.

A membrane (m) Fig 1 — was stretched across the bottom of the box (B). A piece of cork (C) was attached to the centre of the membrane (m) forming a support for the wire W which projected into the water in the glass vessel V. The brass ribbon R was immersed in the water also. Connections were made as in the diagram (Fig 1).

Upon singing into the box, the pitch of the voice was clearly audible from S — which latter was placed in another room. When Mr Watson talked into the box — an indistinct mumbling was heard at S. I could hear a confused muttering sound like speech but could not make out the sense. When Mr Watson counted — I fancied I could perceive the articulations "one, two, three, four, five" — but this may have been fancy — as I knew beforehand what to expect. However that may be I am certain that the inflection of the voice was represented 1, 2, 3, 4, 5.

G.G.H.          Noted March 9th by A.G.B.

---

40

## March 10th 1876

Fig I.

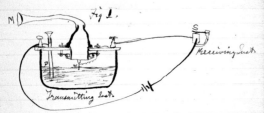

Transmitting Inst.          Receiving Inst.

1. The improved instrument shown in Fig. I was constructed this morning and tried this evening. P is a brass pipe and W the platinum wire M the mouth piece — and S the armature of The Receiving Instrument.

Mr Watson was stationed in one room with the Receiving Instrument. He pressed one ear closely against S and closed his other ear with his hand. The Transmitting Instrument was placed in another room and the doors of both rooms were closed.

I then shouted into M the following sentence: "Mr Watson — Come here — I want to

---

41

see you". To my delight he came and declared that he had heard and understood what I said.

I asked him to repeat the words — He answered "You said 'Mr Watson — come here — I want to see you'." We then changed places and I listened at S while Mr Watson read a few passages from a book into the mouth piece M. It was certainly the case that articulate sounds proceeded from S. The effect was loud but indistinct and muffled. If I had read beforehand the passage given by Mr Watson I should have recognized every word. As it was I could not make out the sense — but an occasional word here and there was quite distinct. I made out "to" and "out" and "further"; and finally the sentence "Mr Bell Do you understand what I say? Do-you-un-der-stand-what-I-say" came quite clearly and intelligibly. No sound was audible when the armature S was removed.

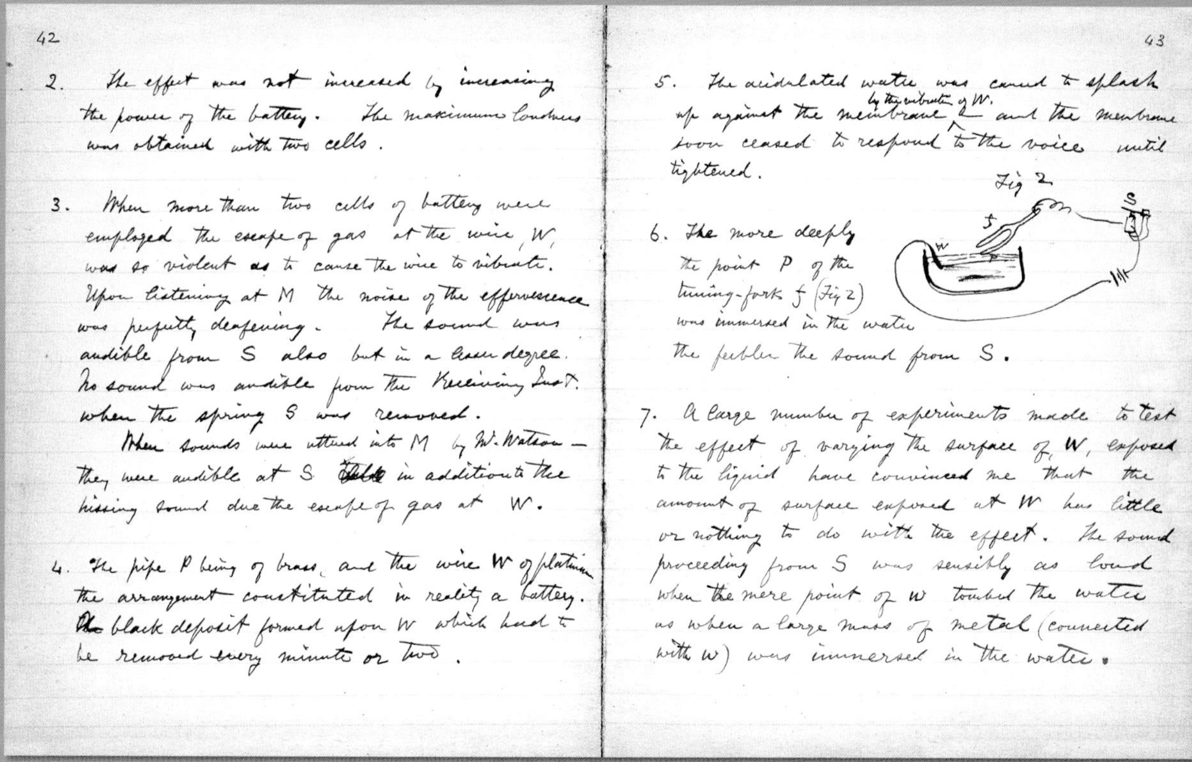

42

2. The effect was not increased by increasing the power of the battery. The maximum loudness was obtained with two cells.

3. When more than two cells of battery were employed the escape of gas at the wire, W, was so violent as to cause the wire to vibrate. Upon listening at M the noise of the effervescence was perfectly deafening. The sound was audible from S also but in a lesser degree. No sound was audible from the Receiving Inst. when the spring S was removed.

When sounds were uttered into M by Mr. Watson — they were audible at S in addition to the hissing sound due the escape of gas at W.

4. The pipe P being of brass, and the wire W of platinum the arrangement constituted in reality a battery. The black deposit formed upon W which had to be removed every minute or two.

43

5. The acidulated water was caused to splash up against the membrane by the vibration of W and the membrane soon ceased to respond to the voice until tightened.

6. The more deeply the point P of the tuning-fork f (Fig 2) was immersed in the water the feebler the sound from S.

Fig 2

7. A large number of experiments made to test the effect of varying the surface of, W, exposed to the liquid have convinced me that the amount of surface exposed at W has little or nothing to do with the effect. The sound proceeding from S was sensibly as loud when the mere point of W touched the water as when a large mass of metal (connected with W) was immersed in the water.

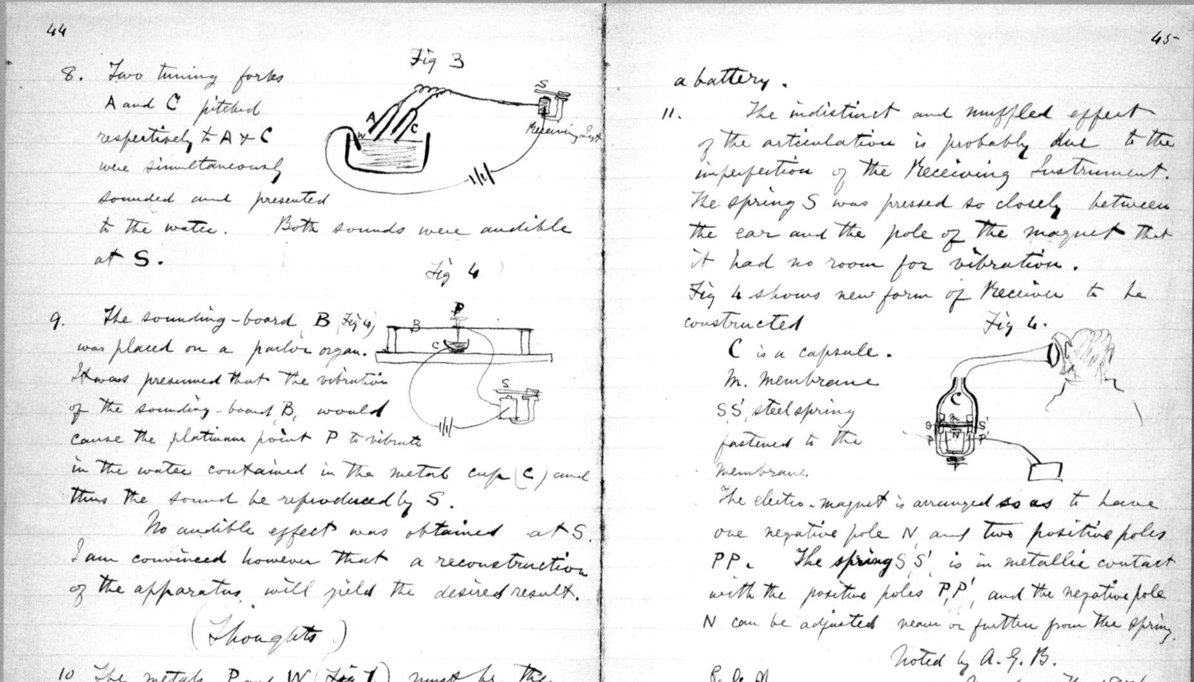

44

8. Two tuning forks A and C pitched respectively, to A & C were simultaneously sounded and presented to the water. Both sounds were audible at S.

Fig 3

9. The sounding-board, B Fig 4, was placed on a parlor organ. It was presumed that the vibration of the sounding-board B, would cause the platinum point P to vibrate in the water contained in the metal cup (C) and thus the sound be reproduced by S.

No audible effect was obtained at S. I am convinced however that a reconstruction of the apparatus will yield the desired result.

Fig 4

(Thoughts)

10. The metals P and W (Fig 1) must be the same to avoid converting the arrangement into

45

a battery.

11. The indistinct and muffled effect of the articulation is probably due to the imperfection of the Receiving Instrument. The spring S was pressed so closely between the ear and the pole of the magnet that it had no room for vibration. Fig 4 shows new form of Receiver to be constructed                    Fig 4.

C is a capsule.
M. membrane
SS' steel spring fastened to the membrane.

The electro-magnet is arranged so as to have one negative pole N and two positive poles PP'. The spring SS' is in metallic contact with the positive poles P,P', and the negative pole N can be adjusted nearer or further from the spring.

Noted by A. G. B.
March 12th 1876

G.G.H.
M.G.H. March 12 /76

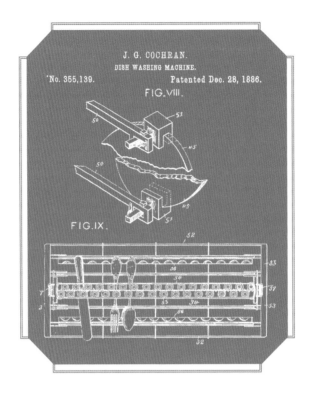

# Domestic Invention

AS ANY STUDY OF INVENTIONS MADE BY WOMEN
SINCE THE MID-NINETEENTH CENTURY PROVES, THERE
IS NO DENYING HOW MUCH THEIR INGENUITY HAS
CONTRIBUTED TO DOMESTIC PROGRESS.

Some of these inventors, like Margaret Knight (1838–1914) started more or less from scratch. Others, like Josephine Garis Cochran (1839–1913), who came up with the idea for an automatic dishwasher, built on the work of others. Though Cochran was not the first person to think of such a device, she was the first to design a reliable, effective model where others had tried and failed before.

Margaret Knight's greatest idea was for a groundbreaking machine that automated the making of flat-bottomed paper bags, cutting out the time-consuming, inefficient business of folding them by hand. She patented her device in 1871, following a long legal battle with Charles Annan, a fraud who had attempted to copy her prototype. Knight went on to file more than 20 patents; according to the *New York Times*, the year before her death she

was still working 20 hours a day, at the age of 70, on her eighty-ninth invention. She had devised her first invention – a shuttle restraint system designed to safeguard workers in cotton mills – when she was only 13.

According to some sources, Ohio-born Josephine Garis Cochran was inspired with the idea for her dishwasher because she hated the chore of washing dishes by hand and did not trust her domestic servants to wash her valuable family china without chipping or breaking it. She may also have been driven by financial necessity; when her husband died suddenly in 1883, he left her with very little money and substantial debt.

What made Cochran's device special was her realization that water pressure, rather than scrubbing, was the key to washing dishes effectively. The contrivance she and George Butters, a young local mechanic, created consisted of a set of wire compartments that fit into a wheel-shaped cage. This was lowered into a copper boiler. A hand-cranked motor turned the wheel and pumped jets of hot, soapy water over the dishes. For rinsing, the machine's user poured hot, clean water over the dishes, which were then left to dry in their own heat.

Cochran patented her dishwasher in 1886. It was the star of the 1893 World Columbian Exposition in Chicago, where it won a prize for "best mechanical construction, durability and adaptation to its line of work". However, the machine proved too costly for anything other than commercial use in restaurants and hotels; most homes of the time were not equipped to supply the amount of hot water Cochran's device required. It was not until the 1950s that dishwashers became popular with the general public.

Dresden housewife Melitta Bentz (1873–1950) was another inventor whose aim was to cut back on the amount of time she spent on domestic chores. In her case, it was the time it took to make a pot of drinkable coffee and the mess that often resulted from its preparation. Before her, making coffee meant wrapping up coffee grounds in a cloth bag, which then had to be steeped in boiling water. The alternative was to pour the water over loose grounds, but the resulting coffee all too often tasted bitter and grainy.

Like many great ideas, the device Bentz came up with to solve the problem was straightforward and simple. She cut a circle from a piece of porous blotting paper, from her eldest son's school notebook, and stuck it in the bottom of an old brass pot into which she had punched some holes. She then poured ground coffee over the blotting paper, followed by hot water. In 1908, Bentz patented her device. What came to be known as the drip coffee-maker had been born.

What drove Anna Bissell (1846–1934) and her husband Melville to invent their carpet sweeper was the mess of sawdust and straw they had to contend with every time they unpacked a delivery at their small Michigan crockery store. Though it was her husband who actually built the device, which he patented in 1876, it was Bissell who put the idea into his head. She also became the sweeper's top salesperson, travelling from town to town to sell them personally and persuading major stores to stock them. When Melville died of pneumonia in 1889, aged only 45, she stepped up to run the company they had founded – becoming the first female CEO in US history – and, by 1899, had expanded it to become the largest organization of its kind in the world. One of her most celebrated customers was Queen Victoria.

Nancy Johnson (1795–1890), who invented the ice-cream freezer in 1843, was not as lucky. Though she patented her idea, she was forced to sell the rights to it because she could not raise the money to finance its commercial manufacture. It was a hand-cranked device, consisting of an outer wooden pail packed with crushed ice and rock salt, and an inner one containing the ice-cream mix that was to be frozen.

Beulah Louise Henry (1887–1973) was given the nickname of "Lady Edison" in the 1920s for her many inventions. She is credited with at least 110 of them, 49 of which she patented over a remarkable 40-year career. She was one of the few women of her time to not only become a professional inventor, but also make a substantial living out of her creations.

Henry patented her first invention in 1912 while she was still a student at a North Carolina college. She was aged just 24; her last patent was granted when she was 82. Throughout her long career, she focused largely on inventions intended to improve everyday life. Notable examples include a parasol with snap-on cloth covers in a variety of colours, a typewriter called the Protograph that produced four copies of an original document without the need for carbon paper and the first bobbin-less sewing machine. As well as doing away with the need to replace bobbins, the machine created a stronger, more durable stitch in half the time.

**OPPOSITE:** Josephine Garis Cochran's original 1886 patent for her "dish washing machine". The bottom figure shows how the cutlery would be held in place whilst cleaned.

**BELOW:** A Grand Rapids Bissel carpet sweeper from 1895.

# Affordable Photography

## IN ITS FIRST 50 YEARS, PHOTOGRAPHY WAS THE PRESERVE OF A RELATIVELY SMALL NUMBER OF PROFESSIONALS AND ENTHUSIASTIC AMATEURS.

It was expensive, time-consuming, awkward and very specialized. All that changed in 1888, when American inventor George Eastman (1854–1932) began selling a cheaper camera, which was also easier to use.

George Eastman was born on a small farm in New York State, USA. When he was five years old, the family moved to the city of Rochester, also in New York. His father died when George was just eight years old, and the family fell on hard times. As a result, George had to leave school aged 13, to find a job. He was keen to learn, though, and was largely self-taught.

Eastman's interest in photography was sparked at age 24 when, while working as a bank clerk, he planned a trip abroad. A colleague suggested he take a record of his trip, so Eastman bought a camera. The camera was a large, unwieldy box, which

had to be mounted on a heavy tripod and instead of film there were individual glass plates that had to be coated with light-sensitive emulsion in situ and held in large plate holders. For outdoor shooting, the plates had to be prepared in a portable tent that doubled as a darkroom.

In 1878, Eastman read about "dry plates", invented in 1871 by the English photographer Richard Leach Maddox. The emulsion was sealed onto the plates with gelatine. These plates could be stored then used whenever desired, making obsolete

**ABOVE:** George Eastman using a 16-mm Cine-Kodak camera.

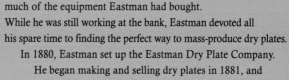

**ABOVE:** The Eastman Dry Plate Company building, in Rochester, New York. Eastman moved to this building in 1883, after the commercial success of his dry plates. Today, Kodak's headquarters are situated at the same address, and the original building has been subsumed into the new.

**RIGHT:** Dry plate camera, 1870s. Photography first took off with the advent of "wet plates" (1850) — glass slides coated in wet, light-sensitive solution. Dry plates were more convenient and afforded shorter exposure times; Eastman's first success was in mass-producing them.

much of the equipment Eastman had bought.

While he was still working at the bank, Eastman devoted all his spare time to finding the perfect way to mass-produce dry plates. In 1880, Eastman set up the Eastman Dry Plate Company. He began making and selling dry plates in 1881, and soon realized that glass could be replaced by a lighter, flexible material. In 1884, he had the idea of making the flexible plate into a roll. A roll holder could be mounted in place of the plate holder inside the camera. His first camera to feature a roll of film, dubbed the "detective camera", became available in 1885. The roll was made of paper, but this was far from ideal since the grain of the paper showed up on the prints. Meanwhile, other people were working on flexible dry plates, too. Several were experimenting with a material called nitrocellulose, also known as celluloid. Eastman began selling celluloid film in 1889.

Eastman's real stroke of genius was his realization that, to be successful, he would need to expand the market for photography, and that would mean, in Eastman's own words, making photography "as convenient as a pencil". To do that, he had to invent a new, smaller, affordable camera. In 1888, the first Kodak camera went on sale. It was an immediate success.

The camera came loaded with a roll able to record 100 photographs. Once a camera's owner had taken the pictures, he or she had only to send the camera to Eastman's company and wait for the pictures and the return of the camera, newly loaded with film. The key to the Kodak's success was changing the perception of photography to something that anyone could do. Eastman had a simple phrase that did just that: "You press the button, we do the rest."

Eastman changed the name of his company to Eastman Kodak, and cornered the market in affordable photography. He never married, nor did he have any children. He was a great philanthropist, giving away large sums of money to universities, hospitals and dental clinics. His last two years were painful as he was suffering from a degenerative bone disease and he took his own life in 1932 by shooting himself in the heart. His suicide note read: "My work is done; why wait?".

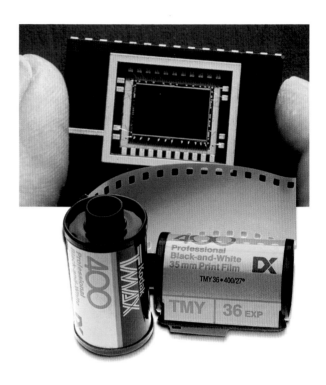

**OPPOSITE:** After its introduction to still photography in 1925, 35mm roll film (below) dominated the market in affordable photography until the introduction of consumer digital cameras in the 1990s. At the heart of a digital camera is a charge coupled device (CCD, above). On the surface of this semiconductor chip are millions of light-sensitive units; each one stores and releases an amount of charge that depends upon the intensity of light that falls on it, and a computer translates those charges into digital information.

**OPPOSITE BELOW:** Girl taking a picture with a Kodak Brownie camera, 1900s. Within a few years, Eastman's Brownie cameras had transformed photography from an expensive, technically-challenging process to, literally, something a child could master and many more people could afford.

## TIMELINE

**1871** Richard Leach Maddox (1816–1902) invents dry gelatin plates, making photography cheaper, more convenient and more portable.

**1878** Eastman invents a machine to mass-produce dry plates.

**1888** The Kodak camera is Eastman's first attempt at simpler, more affordable photography and his first roll-film camera.

**1891** Kodak introduces the daylight-loading film roll – a roll of film in a light-tight container.

**1900** Kodak introduces its first Brownie camera.

**1925** German company Leitz popularizes 35mm film for still cameras when they introduce their first Leica camera.

**1936** Kodak releases its hugely successful Kodachrome colour film on 35mm rolls; the same year, Agfa introduces the even more successful Agfacolor film.

**1948** The Polaroid Corporation introduces the first instant camera, which was invented by American scientist Edwin Land (1909–1991).

**1963** Introduction of the most successful range of low-cost, easy-to-load consumer cameras, Kodak's Instamatic. The Instamatic, which took cartridge film, was immensely successful, introducing a generation to low-cost photography and spawning numerous imitators.

**1969** The charge coupled device (CCD) is invented at Bell Labs, USA.

**1990s** Digital photography reaches the consumer market.

99

## THE BROWNIE

The first camera with mass-market appeal, the Kodak, retailed at $25 (5 shillings in the UK). This was only half what Eastman paid for the first camera he bought, but it was still prohibitively expensive for everyday photography. In 1900, the Eastman Kodak Company introduced the first of its most successful range of cameras: the Brownie. Eastman Kodak made and sold 99 different models of Brownies between 1900 and 1980.

The first Brownie was a cardboard box that contained a roll holder, a roll of film and a lens. On the outside, there was a shutter button and a spool winder. The epitome of simplicity, it sold for just $1 (equivalent to about $20 in 2010), and brought in the era of the "snapshot" – a photograph taken without preparation that can capture a moment in time which would otherwise be lost.

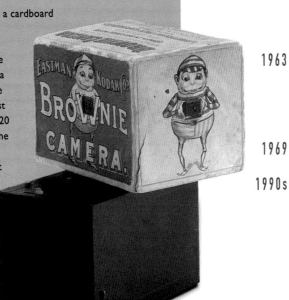

# Alternating Current Motor

ELECTRICITY WAS HAILED IN THE LATE NINETEENTH
CENTURY AS AN EXCITING NEW ALTERNATIVE TO COAL AND
GAS, BUT THE ROAD TO MODERNITY WAS A DIRTY ONE.

Nikola Tesla (1856–1943) was born into a Serbian family in Smiljan, now in the Republic of Croatia but at the time of his birth part of the Austrian Empire. He studied engineering, first in Austria then in Prague, but had to drop out after only a few months due to his father's death. In 1880, he landed a job as a telephone engineer in Budapest.

In 1882, Tesla had a flash of inspiration that resulted in one of his most important inventions: the alternating current (AC) motor. AC is electric current that repeatedly changes the direction it flows along a wire, unlike direct current (DC), which flows in one direction only. Inside Tesla's motor, AC passes through a clever arrangement of coils, producing a rotating magnetic field that spins the rotor.

Also in 1882, American inventor Thomas Edison opened the world's first steam-driven power-generating stations in London and

New York. Both produced DC, which Edison favoured because no AC motors were available, and Edison's light bulbs – the main reason for generating power at the time – did not work well with AC.

**ABOVE:** A demonstration model made by Tesla of an induction motor – perhaps Tesla's most important invention – stripped down to show the coils of wire (stator) surrounding the rotor. Alternating current in the stator creates a rotating magnetic field, which pulls the rotor around.

**OPPOSITE:** Tesla in his Colorado Springs laboratory. To the left is his "magnifying transmitter", which could produce millions of volts. The meandering sparks stretch about 7 metres (23 feet) across the laboratory. The photograph was probably a double exposure with Tesla in one and the sparks in another.

Tesla worked for a year for an Edison subsidiary in France before moving, in 1884, to America. All he had was 4 cents and a letter of recommendation from his boss to Edison himself. Edison gave Tesla a job, and promised him $50,000 if he could improve on Edison's DC generators. Within a year, Tesla had succeeded, but Edison was not forthcoming with the money. Tesla asked for a raise instead, but was again refused, and he resigned.

During the months that followed, Tesla developed a power distribution system based on AC; it was cheaper to install, more efficient and versatile than DC systems and, in 1887 took the precaution of securing patents for his system. American inventor George Westinghouse (1846–1914) was impressed with Tesla's ideas, and in 1888 he gave Tesla a job. There ensued a battle between Edison (DC) and Westinghouse (AC); the famous "War of the Currents" is legendary, but Tesla's system for Westinghouse won out, and his AC motor has driven the wheels of industry ever since.

Around this time, Tesla hit upon two ideas that were to dominate his thinking from then on: the first was the transmission of electric power without wires; the second was wireless transmission of information (radio).

**ABOVE:** Photograph of Nikola Tesla c.1890, just before inventing his famous eponymous Tesla coil.

101

In 1889, Tesla began experimenting with very high-voltage, high-frequency AC (current that oscillates thousands of times every second). Around 1891, he invented the Tesla coil: a kind of transformer that can produce very high voltages. Initially designed to provide wireless power to lights, it played an important role in the development of radio, television and X-ray technology. Meanwhile, Tesla continued his research into wireless broadcasting. Several other inventors were working on the same idea, but Tesla's mastery of high-frequency electricity put him ahead. In 1898, he designed and built the first remotely controlled vehicle: a boat, which he demonstrated to an amazed crowd in Madison Square Garden, New York.

Popular myth blames Mrs O'Leary's cow for kicking over a lantern and sparking the Great Chicago Fire of 1871. Two decades later, in 1893, Chicago's World's Columbian Exhibition (AKA "The World's Fair"), while officially marking 400 years since the arrival of Christopher Columbus in the New World, really celebrated the recovery of its host city. It also augured the slow demise of the old technology that had caused such devastation: the naked flame.

The fairground, nicknamed "The White City", was a dazzling demonstration of electricity, powered using alternating current. It showcased two hundred buildings, canals, lagoons, moving pictures, even the first Ferris Wheel; in short, everything new and exciting about the modern age.

A massive success, it attracted nearly 26 million visitors, many arriving at night just to experience the phenomenon of light by night. Boulevards, interiors and individual exhibits were lit by Westinghouse, who had undercut Edison by 70 cents during the bidding process. Unable to use any of Edison's technology, however, Westinghouse's lamps were short-lived, and a small army of workmen constantly roved the fair replacing bulbs.

Edison did have displays there, as did several other companies, including Tesla himself, who exhibited his "Egg of Columbus", a spinning metal egg demonstrating the rotating magnetic field of a motor. Undoubted draw though the egg may have been, it is likely the masses flocked rather more heavily to witness the Chicago Athletic Association Football team beating West Point in one of the very first night football games.

**BELOW:** Sparks of "artificial lightning" fly from a large tesla coil, Nemesis, built by the Tesla Coil Builders Association, in the USA. Nemesis runs on mains voltage (110 volts in the USA), but produces more than a million volts.

**OPPOSITE:** Wardenclyffe Tower, 57 metres (187 feet) high, with metal pipes pushing 125 metres (400 feet) into the ground. Tesla hoped that electrical oscillations would "shake" the earth and travel through the atmosphere, enabling the worldwide broadcasts of sound and pictures.

NEMESIS

## WAR OF CURRENTS

Nikola Tesla's most important achievement is his design of the power distribution system that has become the standard way of delivering electrical power from generator to consumer. Based on alternating currents, it superseded Thomas Edison's direct current system.

In 1893, the superiority of Tesla's AC system became apparent when the Westinghouse Electrical Company provided impressive electrification of the Chicago World's Fair. That same year, Tesla had the chance to fulfil a childhood dream: to harness the power of the Niagara Falls. He and Westinghouse (above) won the contract to build a power plant there, and their success when the first electricity flowed in 1896 did much to bolster the cause of AC power systems.

In the years that followed, Edison mounted a bitter publicity campaign denouncing AC as dangerous, even going so far as orchestrating public electrocutions of animals and being involved in the development of the first electric chair (which was AC). Despite the campaign, the advantages of Tesla's system guaranteed its success.

## TIMELINE

**1870s** American inventor Thomas Edison and English physicist Joseph Swan (1828–1914) independently develop the first practical light bulbs, creating the need for distributed power supplies.

**1876** Russian electrical engineer Pavel Yablochkov (1847–1894) invents the step-up transformer, later key in the development of alternating current electricity distribution systems.

**1881** World's first power stations open in the UK: at Godalming (hydroelectric) and Holborn Viaduct (steam-powered generators).

**1882** First continuous, reliable electricity supply is provided by Pearl Street Station in New York: electricity for about 3,000 lamps for 59 customers.

**1884** English engineer Charles Algernon Parsons (1854–1931) invents the steam turbine, which begins to replace piston steam engines by 1889, greatly increasing the efficiency and output of power stations.

**1888** American inventor Charles Brush (1849–1929) pioneers generation of electric power from large wind turbines.

**1893** Tesla and George Westinghouse demonstrate the superiority of AC over DC electricity supply, by providing power to the Chicago World's Fair and by producing huge amounts of electric power at Niagara Falls.

**1900–1930** Major programme of electrification creates countrywide "power grids" in industrialized countries.

**1954** Obninsk Nuclear Power Plant in Obninsk, USSR (Russia), is the first to be connected to a power grid.

**1973** Arab oil embargo puts pressure on Western oil supply, leading to crisis and renewed interest in electricity from "renewable" sources, such as solar, wind and hydroelectric power.

**1990s** Growing awareness of global warming caused by carbon dioxide, largely from burning fossil fuels, creates interest once again in renewable, as well as nuclear, power.

103

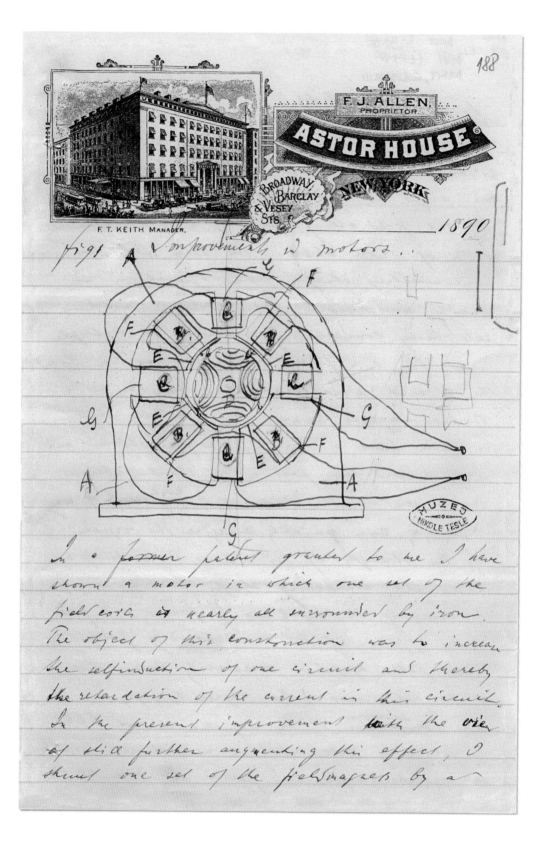

In a former patent granted to me I have shown a motor in which one set of the field coils is nearly all surrounded by iron. The object of this construction was to increase the self-induction of one circuit and thereby the retardation of the current in this circuit. In the present improvement with the view of still further augmenting this effect, I shunt one set of the fieldmagnets by a

# Edison Medal Goes to Nikola Tesla

Nikola Tesla, inventor, was last night adjudged by the American Institute of Electrical Engineers to have contributed the greatest progress to electrical science and electrical engineering during the year 1916. The Edison medal which is awarded each year to the person deemed foremost in this field was presented to him at the annual meeting of the institute, in the Engineering Societies Building, No. 33 West Thirty-ninth street.

The medal is a gold plaque, with the bust of Mr. Edison raised upon one surface. It was presented by H. W. Buck, president of the institute.

The annual report of the Board of Directors was presented, showing a total membership of 8,710. Total revenue during the year was $117,843.68, and total expenses $105,069.21.

**OPPOSITE:** Tesla's drawing and notes for improvements to his alternating current induction motor, on hotel notepaper – Tesla lived in luxurious hotels for much of his life in the USA. These notes relate to US patent 433,701, for which Tesla applied on 26 March 1890.

**ABOVE:** A New York Times article from May 19, 1917, on Tesla's Edison Medal win. Despite declaring him "the person deemed foremost in this field", this was just one year before Tesla had to declare himself bankrupt.

# Motion Pictures

WHILE NO SINGLE PERSON CAN BE CREDITED WITH
INVENTING MOVING PICTURES, TWO FRENCH BROTHERS
STAND OUT FOR THEIR FORESIGHT AND THEIR
IMPORTANT CONTRIBUTIONS.

Using a film camera-projector that they designed, they put on some
of the earliest public film screenings and helped to define cinema.

Auguste (1862–1954) and Louis Lumière (1864–1948) were
born in Besançon, France, where their father Antoine had a
photographic studio. In 1870, they moved to Lyon, and their father
opened a small factory that made photographic plates. In 1882,
Auguste and Louis helped to bring the factory back from the brink
of financial collapse by mechanizing the production of the plates,
and by selling a new type of plate that Louis had invented the
previous year. The firm moved to a larger factory in Montplaisir,
on the outskirts of Lyon, where it employed 300 people.

In 1894, the brothers' father attended a demonstration of
the Kinetoscope, a moving picture peep-show device developed

**ABOVE:** One of the most
important early inventions to come
from Thomas Edison, in 1894, was
the Kinetoscope, the first device
for showing moving pictures.
Edison came up with the idea after
meeting English inventor Eadweard
Muybridge, who pioneered the
photography of movement. The
Kinetoscope and an associated
camera were developed by a British
assistant of Edison, WKL Dickson,
and led to the invention of cinema.

**RIGHT:** One frame from the Lumière Brothers' first film, *La Sortie de l'Usine Lumière à Lyon*, 1895. The film was shot at 16 frames per second and, at that rate, it runs for just under 50 seconds. It features most of the nearly 300 workers – mostly women – walking or cycling out of the factory yard.

**BELOW:** The Lumière Cinématographe – an all-in-one film camera, printer and projector. For shooting, only the camera is needed: the wooden box. The magic-lantern lamphouse – the large black box – contains the light source for projection. The film holder can be seen protruding from the camera-projector box.

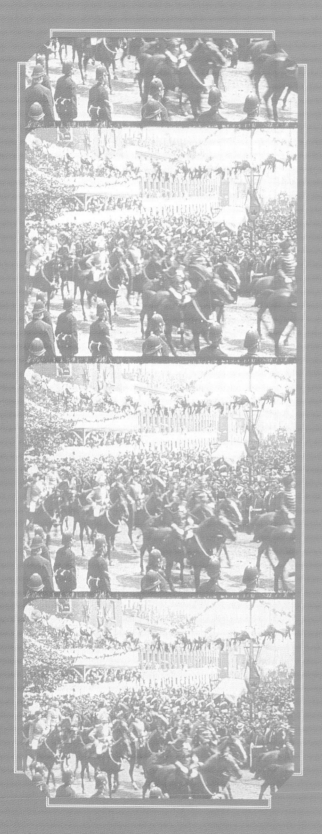

at the laboratory of American inventor Thomas Edison (1847–1931). The Kinetoscope was not a projector – only one person could watch a film at a time – but it was fast becoming popular entertainment. Antoine saw a commercial opportunity and, returning to Lyon, suggested his sons work on producing an apparatus that could record and play back moving images.

Louis, the more technically minded of the two brothers, designed the camera-projector, while Auguste designed the housing for the light source. Louis developed the film transport mechanism, inspired by a similar device in sewing machines, which allowed each frame of the film to stop momentarily behind the lens.

The Lumière brothers patented their camera-projector, the Cinématographe, in February 1895. Louis shot their first film, which was called *La Sortie de l'Usine Lumière à Lyon* (*Workers Leaving the Lumière Factory in Lyon*), and the pair showed the film to the Société d'Encouragement de l'Industrie Nationale, in Paris in March 1895, the first public screening of a film using their projector.

After several other screenings in France, their father arranged for the first performances to a paying audience. Ten films were shown 20 times a day. The opening night, at the Salon Indien – the empty basement of the Grand Café in Paris – was in December 1895. Auguste and Louis did not attend the first day, because they felt the technology still needed more work.

After a slow start, the shows became a great success. In 1896, the Lumière brothers sent their agents abroad, demonstrating their Cinématographe and arousing great interest. They also ordered 200 or so of the camera-projectors to be constructed, and opened agencies in several countries to sell them. The Lumière franchise was very successful, but they refused to sell their devices to anyone except through their own agents.

By 1897, Thomas Edison had developed a system of sprocket holes that was incompatible with the Cinématographe and that was quickly becoming the standard in a rapidly developing industry. By 1905, Edison's system would predominate and the Lumière brothers would leave the film business.

Auguste's interests turned to chemistry and medicine. In 1910, he founded a laboratory in Lyon, where his 150 staff carried out research into cancer and other diseases. Auguste invented a dressing for burns, called *tulle gras*, which is still used today, and pioneered the use of film in surgery, which helped generations of medical students. Meanwhile, in the early 1900s, Louis demonstrated a sequence shot on a new, wider-format film, and later experimented with panoramic and stereoscopic (3D) films.

In 1904, the Lumière brothers perfected a colour photography system called Autochrome; they had been working on colour photography since the early 1890s. Autochrome was the most important colour photographic process until colour film became available in the 1930s.

**OPPOSITE:** The Lumières' film of Queen Victoria's Diamond Jubilee procession in London, 1897. The film's circular sprocket holes are characteristic of the Lumières' system. Other early filmmakers used 35mm film with rectangular Edison perforations, which became the industry standard. Later Cinématographes could run "standard" 35mm film

**RIGHT:** Poster, late 1890s, advertising the Lumière brothers' Cinématographe. The poster shows three of a film's 900 frames – each one slightly different. Each frame is a photograph taken about one-sixth of a second after the next, since the Lumière brothers shot their films at 16 frames per second.

## PIONEERS OF MOTION PICTURES

The Lumière brothers may have pioneered cinema, but they were not the first to make moving-picture films. Many inventors, scientists and photographers were experimenting with moving pictures several years before the Lumières. One of the first people to capture realistic movement on film was French inventor Louis Le Prince (1841–1890). Le Prince made his first successful film in October 1888. This was a sequence shot in his father-in-law's garden at Roundhay in Leeds,

England, showing his son, his in-laws and a family friend.

The Lumières were not the first to project films to a paying audience, either. Projection and a paying audience form the definition of cinema. That honour of the first cinema performance goes to American brothers Grey and Otway Latham, who projected their films in New York in April 1895. But the "projector" they were using was simply a modified Edison Kinetoscope, and the results were not very good.

**ABOVE:** Colour photograph, c.1910, taken with the Lumières' Autochrome system. When shooting, a glass slide coated with randomly scattered red-, green- and blue-pigmented starch grains was held in front of the (black- and-white) film; the same slide was required for viewing.

**BELOW:** Auguste looking down the microscrope in his laboratory, assisted by his brother Louis, in the 1930s.

## TIMELINE

**1820s–1860** Toys such as the thaumatrope (1824), the zoetrope (1833) and the flick book (1860) give the illusion of movement by presenting the brain with a quick succession of images.

**1877** English photographer Eadweard Muybridge (1830–1904) uses a line of cameras to capture the motion of a galloping horse as a sequence of individual pictures.

**1879** Muybridge demonstrates his Zoopraxi-scope – a device for projecting very short sequences of moving pictures.

**1888** Louis Le Prince shoots his first film, the *Roundhay Garden Scene*.

**1889** English inventor William Friese-Greene (1855–1921) receives a patent for his chronophotographic camera, designed to capture ten frames per second on perforated film.

**1891** American inventor Thomas Edison (1847–1931) patents a moving-picture camera (Kinetograph) and display device (Kinetoscope). The devices use 35mm film, which later becomes the industry standard.

**1892** French showman Charles-Émile Renaud (1844–1918) performs his *Théâtre Optique*, a theatrical show featuring the first animation films, using his own hand-drawn images.

**1895** The Lumière Brothers patent their Cinématographe and give its first public projections.

**1895** Grey (1867–1907) and Ottway (1868–1906) Latham become the first to project films to a paying audience.

**1908** Albert Smith (1875–1958) and J Stuart Blackton (1875–1941) produce the first stop-motion animation film, *The Humpty Dumpty Circus.*

**1908** The first commercial colour moving-picture films are made using a British system called Kinemacolor; the more successful American Technicolor process debuts in 1922.

**1927** *The Jazz Singer*, featuring singer Al Jolson, is the first picture with synchronized speech – heralding the age of the "talkies".

**1995** The release of the first film made solely using computer-generated imagery, *Toy Story.*

111

# Powered Flight

## AT THE DAWN OF THE TWENTIETH CENTURY, TWO BROTHERS FROM A SMALL TOWN IN THE USA BECAME THE FIRST TO ACHIEVE SUSTAINED, POWERED FLIGHT.

The key to their success was the combination of their inventive, mechanical skill with the application of scientific principles to flight. Moreover, they learned to become pilots in a gradual, thoughtful way, rather than risking everything on one short trial, like so many other pioneers.

Wilbur (1867–1912) and Orville (1871–1948) Wright grew up in Dayton, Ohio, in a family with seven children (although two died in childhood). They were mechanically-minded from an early age: in 1886, they built their own lathe; in 1888, they built a printing press, which they used to produce their own local paper; and in 1892, they opened a bicycle repair shop. They used the profits of the shop to finance their efforts in aviation.

The dream of human flight stretches back to antiquity, but it was only in the late eighteenth century that people finally made it into the air, by courtesy of "lighter-than-air" balloons. In the nineteenth century, scientists and inventors began giving serious consideration to the problem of "heavier-than-air" flight. Providing power was problematic, since steam engines were large and very heavy. During

**ABOVE:** The Wright Flyer's first flight. Orville is piloting (lying down, to reduce drag), Wilbur beside the wing tip. The rudders are at the rear, as are the two counter-rotating propellers, which are blurred out. The Flyer was launched from a wooden track.

112

## OTTO LILIENTHAL (1848–1896)

It was Wilbur who was first struck by the desire to build a powered flying machine, after reading a magazine article about Otto Lilienthal, a German gliding pioneer. Lilienthal realized that to develop successful flying machines, any inventor needed to understand the scientific principles behind flight but also needed first-hand experience of flying. The Wright brothers took the same approach, and paid tribute to Lilienthal as their inspiration.

Lilienthal began his quest to fly by studying birds, and then carried out a huge amount of research into aerodynamics. In the 1890s, he made nearly 2,000 flights, mostly from an artificial hill he built near Berlin, in gliders he had designed and constructed. During what was to be his last flight, a gust of wind made him stall at an altitude of 15 metres (50 feet). He crashed to the ground and died from his injuries the next day. According to legend, his last words were, "Sacrifices must be made."

**BELOW:** Replica of the engine that powered the Wright Flyer. The engine was built in 1903 by Wilbur and Orville's bicycle-shop mechanic, Charlie Taylor (1868–1956). It was relatively light, thanks to the fact that the cylinder block was cast in aluminium.

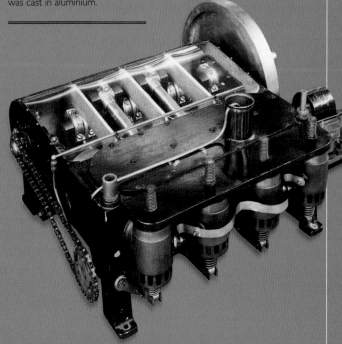

114

the 1880s and 1890s people flew in unpowered gliders and kites. In 1899, the Wright brothers built a large box kite. Wilbur hit on the idea that by twisting the box shape, it would be possible to change the airflow over the wings and make the kite bank and turn. He called this "wing warping", and it would be crucial to the brothers' later success.

After the kite performed well, the brothers decided to build full-size, piloted gliders, with wing warping effected via control cables. They constructed their first glider in 1900, and also added a front "wing" called an elevator, for pitch control. They chose the open area on the coast, near the tiny fishing village of Kitty Hawk in North Carolina, for its steady on-shore winds. First they flew the glider tethered like a kite, moving to the nearby Kill Devil Hills for actual flights. During 1901 and 1902, Wilbur and Orville built and tested two more gliders, and they also carried out hundreds of experiments in a homemade wind tunnel in their bicycle shop back in Dayton. By analysis and practical trials the brothers became the first to realize that controlling an aircraft required the banking control (wing warping or aileron), rudder and elevator all to be used continuously in combination. They were now ready to make a powered version of their flying machine. For driving the aeroplane, they designed and

built large wooden propellers and, with a colleague in the bicycle shop, made a purpose-built, lightweight, powerful engine.

In December 1903, at Kill Devil Hills, the Wright brothers were ready to put all their ideas, experiments and calculations to the test. The first successful flights took place on December 17. There were four flights that day, two by each brother. The first, with Orville piloting, lasted just 12 seconds and covered 37 metres (120 feet). The final flight of the day, with Wilbur as pilot, lasted 59 seconds and covered 260 metres (852 feet). By 1905, the Wright brothers' flying

---

**TOP LEFT**: A posed photograph of the Wrights, with Orville on the left and his brother, Wilbur, on the right. Three years later they would found The Wright Company.

**ABOVE:** Certificate granting US patent 821,393 to the Wright brothers, for a "flying machine". The patent focused on the construction of their aeroplane and its control mechanisms. The brothers applied for it in 1903, and were awarded it on 22 May 1906.

machines were routinely staying in the air for several minutes at a time, taking off, landing, and manouevering with ease. At first, the world was slow to recognize the Wrights' achievement, despite the fact that there were several witnesses on the day. This was partly because the media and the public were unwilling to believe that the age-old dream of flight had finally come true, but also because the brothers became secretive about their work, hoping to sell their invention to a government or large corporation.

Wilbur and Orville were awarded a patent in 1906, for a "Flying Machine". Three years later they founded The Wright Company, to take advantage of the patent. Unfortunately, Wilbur died within three years, from typhoid. Orville went on to become a long-time advisor to the US Government's National Advisory Committee for Aeronautics, and was able to appreciate the incredibly rapid developments in aviation that took place within a few decades of those first flights.

**BELOW:** The Wright brothers' wind tunnel, which they built in their bicycle workshop in Dayton, Ohio. They used the wind tunnel to test wing designs – in particular the appropriate wing camber – to compile the first accurate tables of lift and drag forces on wings and understand how the lift force moves back or forward as the wing tilts, affecting control.

# TIMELINE

**1783** French inventor brothers Joseph-Michel (1740–1810) and Jacques-Étienne Montgolfier (1745–1799) demonstrate their hot-air balloon, which achieves the world's first piloted ascent.

**1799** English engineer George Cayley (1773–1857) experiments with lift and establishes the basic form of fixed-wing, heavier-than-air craft.

**1852** French engineer Henri Giffard (1825–1882) builds the first powered airship.

**1896** American scientist and engineer Samuel Pierpont Langley (1834–1906) achieves the first sustained powered flight, with his un-piloted, steam-powered model Aerodrome Number 5.

**1890s** Otto Lilienthal makes nearly 2,000 flights in gliders of his own design and construction.

**1903** The Wright brothers become the first to fly sustained powered flights, in their Wright Flyer.

**1919** German airline Deutsche Luft Reederei becomes the first airline to run scheduled daily flights.

**1930s** English engineer Frank Whittle (1907–1996) and German engineer Hans von Ohain (1911–1998) independently develop the jet engine.

**1930s** The first viable helicopters take flight, in France, Germany, Russia and the USA.

**1947** Charles E Yeager pilots Bell X-1, the first aircraft to exceed the speed of sound in level flight.

**1949** A US Air Force B-50 Superfortress, *Lucky Lady II*, makes the first non-stop flight around the world.

115

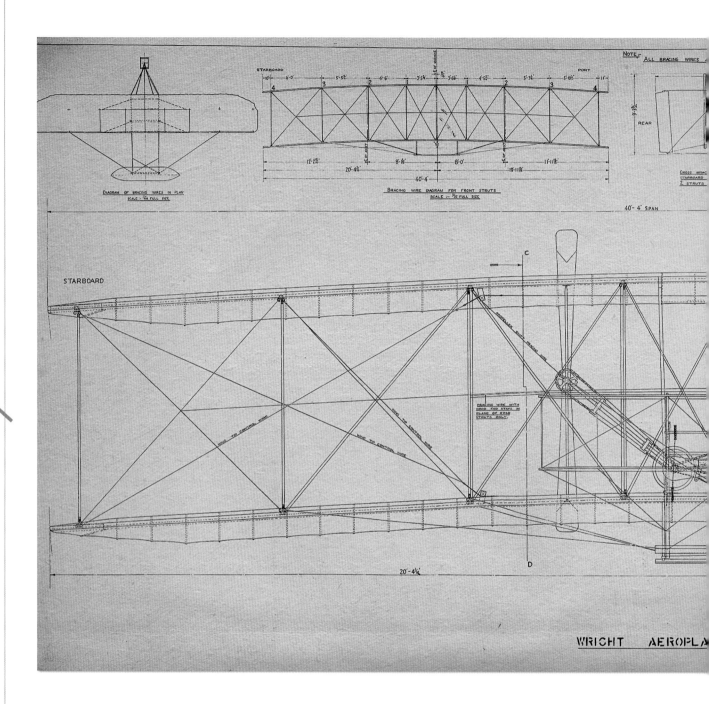

**ABOVE:** Front elevation drawing of the Wright Flyer. The engine was mounted slightly left of centre in this view, and it powered the two propellers (at the back of the aircraft) via sprockets and chains – similar to how pedals power a bicycle.

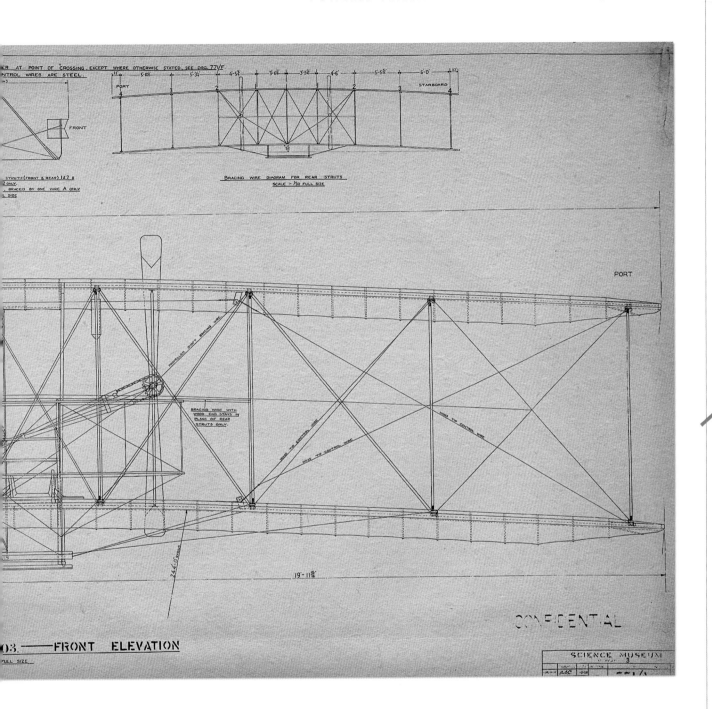

ER AT POINT OF CROSSING EXCEPT WHERE OTHERWISE STATED, SEE DRG. 771/F.
NTROL WIRES ARE STEEL

FRONT

PORT                                                  STARBOARD

BRACING WIRE DIAGRAM FOR REAR STRUTS
SCALE 1:32 FULL SIZE

STRUTS (FRONT & REAR) 1 & 2 &
2 ONLY.
BRACED BY ONE WIRE A ONLY
L SIZE

PORT

BRACING WIRE WITH
WOOD END STAYS IN
PLANE OF REAR
STRUTS ONLY.

19'-11⅝"

03.——FRONT ELEVATION

FULL SIZE

117

# Synthetics

IN 1946, WHEN STEPHANIE KWOLEK STARTED WORKING
AS A YOUNG SCIENCE GRADUATE FOR DUPONT IN ITS
TEXTILE FIBRE LABORATORY, OPPORTUNITIES FOR FEMALE
CHEMISTS WERE FEW AND FAR BETWEEN.

Pennsylvania-born chemist Stephanie Kwolek's (1923–2014) original plan was to work for DuPont temporarily until she had raised the money she needed to fund her way through medical school.

As it transpired, Kwolek found the work so fascinating that she stayed with the company for the whole of her scientific career, which lasted for more than 40 years. "The work was so interesting and so challenging," Kwolek told a student audience in 1986. "I loved to solve problems and it was a constant learning process. Every day, there was something new, a new challenge, and I loved that... I was so interested in chemistry and research that I totally forgot about medicine."

Kwolek's interest in science had been originally fostered by her father, who encouraged her exploration of the wooded areas near her home. At DuPont, her research focused on polymers, the so-called long molecules that are created by joining smaller molecules together to form a chain. Initially, what she was looking for was a strong, lightweight polymer fibre that could be used in place of the steel wires that were an important constituent of the car tyres of the time. Her work, with what

**ABOVE:** Katherine Burr Blodgett demonstrating her equipment in the General Electric Laboratory, 1938.

are scientifically named "aramids", led her to the discovery of a liquid crystalline polymer solution that, when spun into fibre, proved to be ounce for ounce five times stronger than steel, yet lighter than fibreglass.

This was in 1965. The discovery led to the creation of a multi-billion dollar industry, following DuPont's release of what it trademarked as "Kevlar" in 1971. It cost the company an estimated $500 million to develop the new synthetic to the point where it could be utilized commercially, but the investment soon proved value for money. As well as being exceptionally strong and stiff, Kevlar proved to be hard-wearing and fire and corrosion resistant. One of its first major uses was as the main component in the production of bulletproof vests and body armour. Its success won Kwolek renown as the person who invented a fibre that saved thousands of lives; at the time of her death, a DuPont spokesperson estimated that Kevlar had saved the lives of 3,000 police officers in the US alone.

Today, in its numerous variants, Kevlar has more than 200 applications, ranging from the protection of undersea fibreoptic cables, the construction of suspension bridges, rockets and the International Space Station, to being used in boat hulls, tennis racquets, skis, protective clothing for athletes and even as a coating for the humble frying pan. Although Kwolek did not make any money out of Kevlar personally – she assigned all her patent royalties to DuPont – she went on to make further important discoveries. In all, she was the sole holder of seven patents, while her name also appeared on 16 more.

Katharine Burr Blodgett (1898–1979) was another trailblazer. Not only was she the first female scientist to be hired by General Electric – she started work there in 1918 – but her invention of so-called "invisible" glass transformed the world. Today, the non-reflective coatings she devised are used on everything from windscreens and camera lenses to eyeglasses and computer screens.

At General Electric, Blodgett was mentored by Nobel Prize-winning chemist and physicist Irving Langmuir (1881–1957), who was investigating how substances stick together at the molecular level. He had got as far as creating a film on the surface of water that was just one molecule thick, before asking Blodgett to take the research further and see if she could find practical applications for these findings.

Blodgett lived up to Langmuir's expectations, taking the research a vital step further by discovering how to add layers to the films. This involved dipping a metal plate into water that was covered with a layer of oil. Numerous repetitions of the process meant that Blodgett could literally stack layers of oil on the plate. She soon found herself able to control the exact thickness of the films, down to the molecule.

The real breakthrough came when Blodgett tried adding layers of film to both sides of a sheet of glass until the visible light reflected by the layers cancelled out the light reflected by the glass itself. She went on to invent ways of measuring the thickness of these coatings to a millionth of an inch, using the principle that different thicknesses of coating are differently coloured.

Invisible glass had been born. One of its first practical applications was in the movie industry, where non-reflective lenses transformed the results when films were shot and projected. During the Second World War, non-reflective glass was an essential component in aircraft spy cameras and submarine periscopes.

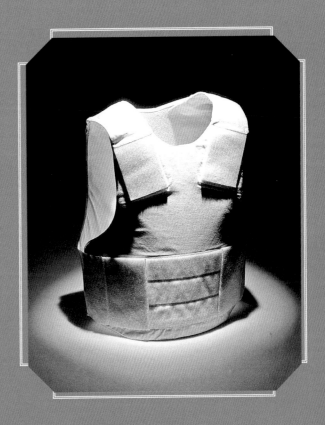

**BELOW:** A modern bulletproof Kevlar vest used by law enforcement and members of many special forces worldwide.

119

# Radio

## THE EARLY HISTORY OF RADIO IS COMPLEX, AND CREDIT IS DUE TO DOZENS OF IMPORTANT PIONEERS.

One of the most important and successful was Italian inventor Guglielmo Marconi, who helped bring radio into everyday use. Guglielmo Marconi was born in Bologna, Italy to an Italian father and an Irish mother. From an early age, he took an interest in science and was particularly interested in electricity. In late 1894, Marconi became aware of the experiments of the German physicist Heinrich Hertz (1857–1894), who had succeeded in proving the existence of radio waves during the late 1880s.

Hertz produced radio waves by sending a rapidly alternating current up and down an antenna, and detected the waves up to 20 metres (65 feet) away. Marconi also read about a demonstration that English physicist Oliver Lodge (1851–1940) had recently performed. Lodge sent Morse-code messages wirelessly, using the "Hertzian" waves. At the time, telegraph messages in Morse code could only be sent as electric pulses along wires, and Marconi was excited at the prospect of "wireless telegraphy".

Marconi decided to carry out experiments of his own, with the aim of making wireless telegraphy a useful, practical technology. He set up a laboratory in the attic room of his family home, and assembled the necessary components. He was soon sending and receiving Morse code wirelessly over increasingly large distances: first across the room, then down a corridor, then outside, across fields. In the summer of 1895, Marconi transmitted a message over

**ABOVE:** Marconi in 1922, in the radio room of his yacht *Elettra*. Here he picked up radio signals he believed had come from planet Mars. In fact, they were probably "whistlers" – very low-frequency waves produced by lightning.

nearly 2 kilometres (1.2 miles), and in 1896 patented his system. On being refused funding by the Italian government, he decided to travel to Britain to seek interest there.

Following a series of impressive demonstrations during 1897, Marconi garnered the support of the Post Office, which was in charge of Britain's telegraph system at the time. In that year, he formed the Wireless Telegraph & Signal Company to expand his work. In the following few years, he sent messages over ever greater distances and, notably, between ships and from ship to shore. In 1900, Marconi decided to try extending the range of his transmissions yet further: across the Atlantic Ocean. In 1901, he created a worldwide sensation when he announced the successful transmission of a Morse code letter "S" (three short bursts of radio) from Poldhu, in Cornwall, England to St John's, Newfoundland (then a British colony, now in Canada). After suggestions that he had faked the transmission, he carried out another, carefully monitored experiment the following year. Aboard a ship close to the Canadian coast, he received signals from Cornwall more than 3,200 kilometres (2,000 miles) away.

During the years that followed, Marconi made several important improvements to his system of radio transmission, and in 1907 he instigated the first commercial trans-Atlantic radio service. He found fame again when the British ocean liner RMS *Titanic* hit an iceberg and sank in 1912. A Marconi-radio operator aboard the sinking ship managed to broadcast radio distress signals and summon help from nearby ships.

**ABOVE:** Illustration of the receiving end of Marconi's first trans-Atlantic transmission in 1901. A kite was used to lift the antenna into the air at St Johns, Newfoundland, after a heavy storm destroyed the original fixed antenna at Cape Cod, Massachusetts.

**LEFT:** Sailors on board ship, reading a "marconigram", in the early 1900s. Just as a telegram was a physical record of a Morse code message sent via telegraph wires, a marconigram was a record – on paper tape – of Morse-code message received wirelessly via radio.

During the 1920s, Marconi experimented with much higher-frequency radio waves. These "short waves" can be focused by a curved reflector behind the transmitter, like the parabolic dishes used to receive satellite communications. This arrangement made radio more efficient and less power-hungry, since the waves were concentrated into a beam and not radiating in all directions. By this time, radio operators, including Marconi, were transmitting not only Morse code, but also speech, music and audio signals. In 1931, he experimented with even higher-frequency, shorter-wavelength radio waves – microwaves – and a year later, he installed a beamed, microwave radio-telephone system between the Vatican and the Pope's summer residence. Much of today's telecommunications infrastructure is built on microwave beams like this.

Marconi did not invent radio, but he did make several important improvements to it, and his determination to turn a complicated laboratory curiosity into something useful and commercially successful helped make the world feel a bit smaller. In 1909, he received the Nobel Prize for Physics, for his contributions to wireless telegraphy, and in 1930, he became president of the Royal Italian Academy.

## LEE DE FOREST
## (1873–1961)

In the early 1900s, radio communication could only be made using wireless telegraphy – sending Morse-code messages as on-and-off pulses of radio waves. That changed with the introduction of audio broadcasting; regular broadcasts began in 1920. One of the most important technologies involved in the development of audio broadcasting was the Audion, invented in 1906 by American electronics engineer Lee de Forest .

The Audion was an early example of a "valve", which found myriad uses in the developing field of electronics. In radio and television broadcasting, it enabled the construction of all-electronic "oscillators", which produced radio waves of any frequency to order. From the 1920s until the 1960s, radio and television sets used valves for amplification. Eventually, they were replaced by the smaller, less power-hungry transistor, invented in 1947.

**OPPOSITE:** "Marconiphone" amplifier from around 1925, with valves – the developments of de Forest's Audion (see above). Marconi formed the Marconiphone Company in 1922, to manufacture radio sets for domestic use as well as amplifiers like this one, which made it possible to listen without headphones.

**OPPOSITE TOP:** Batteries and tuning coils at Marconi's South Wellfleet station, Massachusetts. From here, in 1903, Marconi sent a message from US President Theodore Roosevelt to King Edward, in London – a distance of more than 5,000 kilometres (3,000 miles).

## TIMELINE

**1888**
Heinrich Hertz completes his experiments in which he proves the existence of radio waves.

**1893–1895**
Three scientists give public demonstrations of wireless communication: Serbian-American genius Nikola Tesla, Oliver Lodge and Bengali polymath Jagadish Chandra Bose (1858–1937).

**1901**
At St John's, Newfoundland, Marconi receives the letter "S" in Morse code, across the Atlantic Ocean from Cornwall, England.

**1904**
English physicist John Ambrose Fleming (1849–1945) invents the thermionic diode valve, the first electronic component, which plays an important role in the development of radio.

**1906**
Lee de Forest adds a "control grid" to Fleming's valve, creating the Audion, a device that can amplify.

**1906**
Canadian inventor Reginald Fessenden (1866–1932) becomes the first to transmit sound by radio – music and a reading from The Bible – which is heard by ships' radio operators.

**1920**
The first regular radio entertainment broadcasts begin, in Argentina, pioneered by entrepreneur Enrique Susini (1891–1972).

**1933**
American electrical engineer Edwin Howard Armstrong (1890–1954) invents frequency modulation (FM), which allows radio broadcasts to be clearer and interference-free.

**1940**
English physicist Harry Boot (1917–1983) and John Randall (1905–1984) invent the cavity magnetron, which generates high-frequency, short-wavelength radio waves: microwaves.

**1950s**
Transistors rapidly take over from valves in radio sets. Transistor radios are cheaper and more portable.

**1999**
The first commercial digital radio broadcasts.

123

# The Haber-Bosch Process

## THERE IS ONE TWENTIETH-CENTURY INVENTION THAT ARGUABLY CHANGED THE WORLD MORE PROFOUNDLY THAN ANY OTHER.

It is not a machine or a device, but an industrial process. The manufacture of ammonia, perfected by German chemist Carl Bosch (1874–1940), enabled the production of fertilizers and explosives on a completely unprecedented scale, resulting in a meteoric rise in population and unlimited explosive capacity in two world wars.

Carl Bosch was born in Cologne, Germany. He studied mechanical engineering and metallurgy at Charlottenburg Technical University. In 1896, he began studying chemistry, at the University of Leipzig. Three years later, Bosch joined Germany's most successful chemical company, in Ludwigshafen. At the time, the company's name was Badische Anilin- & Soda-Fabrik; nowadays, the name is simply BASF.

At first, Bosch worked on synthetic dyes, but in 1905 he turned his attention to a major question of the day: how to "fix" atmospheric nitrogen into chemical compounds. This seemingly esoteric issue was actually of immense global

**ABOVE:** The world's first ammonia synthesis plant, at Oppau, near BASF's headquarters in Ludwigshafen, Germany. In its early years, the plant produced more than 7,000 tonnes of ammonia, made into 36,000 tonnes of ammonium sulphate. The four large towers are ammonia storage silos.

**BELOW:** German Nobel Prize-winning chemist Carl Bosch, best known for scaling up the production of ammonia to an industrial level, enabling the manufacture of artificial fertilizers.

**RIGHT:** A German First World War biplane dropping a bomb. At the time, the manufacture of explosives depended upon a plentiful source of nitrogen-rich compounds. Bosch's process for the manufacture of ammonia helped Germany meet the demand and sustain its war effort.

In the early years of the twentieth century, a growing threat of war led to further increases in the demand for nitrogen compounds.

As an element, nitrogen is notoriously unreactive. That is why it makes up nearly 80 per cent of the atmosphere. From the 1890s, chemists had tried in vain to find an efficient, high-yield process to fix nitrogen from the air to make fertilizers and explosives. Then, in 1905, German chemist Fritz Haber (1868–1934) reported that he had produced small amounts of ammonia from nitrogen gas ($N_2$) and hydrogen gas ($H_2$). Haber's process required high temperature, high pressure and a catalyst – a chemical that speeds up a reaction, while remaining unchanged, or a chemical that lowers the energy needed for a reaction to take place, therefore speeding it up. Haber was working under contract to BASF and, by 1909, he had produced an impressive yield of ammonia in his laboratory. In that year, BASF gave Bosch the task of scaling up Haber's reaction for use on an industrial scale.

significance. Scientists in the nineteenth century had realized that nitrogen-rich compounds made very effective fertilizers. In particular, huge deposits of guano (fossilized bird excrement) and saltpetre (potassium nitrate, $KNO_3$) had helped to sustain an ever-expanding world population. In 1898, English chemist William Crookes (1832–1919) delivered a lecture to the British Association entitled "The Wheat Problem", in which he noted that these deposits were dwindling. Crookes suggested that the world could face major famines by the 1920s. In addition, nitrogen compounds were an essential ingredient in explosives.

Bosch developed a reaction vessel that could withstand the high temperatures and pressures that were necessary: a double-walled chamber that was safer and more efficient than Haber's system. He carried out nearly 20,000 experiments before he found a more suitable catalyst than the expensive osmium and uranium Haber had used. Bosch also worked out the best ways to obtain large quantities of hydrogen – by passing steam over red-hot coke – and nitrogen, from the air. He patented his results in 1910, and by 1911, BASF had begun producing ammonia in large quantities. The company opened the world's first dedicated ammonia plant, in Oppau, a suburb of Ludwigshafen, just two years later. The ammonia was used to make artificial fertilizers in huge quantities. When the First World War began in 1914, however, the German government was faced with a shortage of ammunition, and the output of the Oppau plant was used to produce explosives instead. Without the Haber-Bosch process, the war would probably not have lasted as long as it did; Britain had blockaded Germany's imports of saltpetre, which Germany had relied upon to make explosives.

Bosch's intensive work and his insight into chemistry and engineering helped to lay the foundations of large-scale, high-pressure processes – which, in turn, underpin much of the modern chemical industry. In 1931, he was awarded the Nobel Prize for Chemistry. Today, nearly 200 million tonnes of synthetic nitrogen fertilizers are produced worldwide every year – several tonnes every second – using the Haber-Bosch process.

**ABOVE RIGHT:** Synthetic ammonia fertilizer factory, 1920s. Before Bosch developed his industrial process, only bacteria, living in the soil or in water, could "fix" nitrogen from the air in these quantities. The production of synthetic nitrogen-based fertilizers enabled the world to avoid mass starvation.

**MIDDLE RIGHT:** A tractor spraying nitrogen-rich artificial fertilizers on a rice paddy field in Spain. In the twentieth century, the availability of artificial fertilizers and the mechanization of farming equipment led to the rise of intensive agriculture, a system with high-energy input and dramatically increased crop yields.

**RIGHT:** German chemist Fritz Haber, photographed in 1918, the year he won the Nobel Prize for Chemistry. Haber developed the reaction that produces ammonia from nitrogen and hydrogen gases, which Bosch successfully scaled up in 1910; the resulting technique is today called the Haber–Bosch Process.

## FERTILIZERS

Carl Bosch made it possible to produce huge quantities of ammonia, much of which is made into nitrogen-rich ammonium nitrate ($NH_4NO_3$) fertilizer. Careful estimates suggest that synthetic fertilizers feed about half of the world's population. Plants rely upon nitrogen compounds for building proteins and DNA (deoxyribonucleic acid). In nature, nitrates come from decaying plant and animal matter and from certain bacteria, which fix nitrogen from the air.

Bosch's lasting legacy is double-edged, however. Artificial fertilizers have saved millions from starvation, but the huge increases in population they allowed, from nearly 1.8 billion in 1910 to nearly 7 billion a century later, have put a strain on the world's resources. Their manufacture accounts for about one per cent of the world's total energy consumption and their use causes pollution; agricultural run-off in particular creates "harmful algal blooms" in lakes and estuaries due to the extra nitrogen.

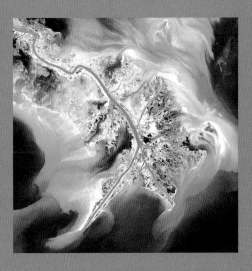

**ABOVE:** The heavy use of artificial fertilizers causes dead zones, like this one in the Gulf of Mexico. Dead zones form in lakes and coastal seas as agricultural run-off finds its way into water courses, causing a proliferation of algae (as a result of the extra nitrogen), which starve other organisms of oxygen.

## TIMELINE

**1854** — English chemist William Perkin (1838–1907) discovers the first synthetic dye, mauvine, from coal tar. Soon a whole industry springs up, based on coal tar derivatives.

**1865** — Swedish chemist Alfred Nobel (1833–1896) builds a factory near Hamburg, Germany, producing large quantities of nitroglycerin as explosives.

**1907** — Belgian chemist Leo Baekeland (1863–1944) develops Bakelite, the first plastic to be manufactured on an industrial scale.

**1907** — Fritz Haber becomes the first person to produce significant quantities of ammonia in the laboratory, from nitrogen and hydrogen.

**1910** — Working at BASF, Carl Bosch scales up Haber's reaction to an industrial scale, heralding a new era of artificial fertilizers.

**1916** — German chemist Fritz Günther (1877–1957) produces Nekal, the first synthetic detergent, to alleviate soap shortages at the end of the First World War.

**1920s** — Crude oil – petroleum – begins to take over from coal tar as the main raw material in the chemical industry. Petrochemicals form a huge range of different products, including plastics, flavourings, preservatives, soaps, solvents and pesticides.

**1933** — British company ICI (Imperial Chemical Industries) begins production of polythene, the most widely used plastic.

**1939** — Swiss chemist Paul Herman Müller (1899–1965) discovers the insecticidal properties of DDT (dichlorodiphenyltrichloroethane), which is then used in large quantities to fight malaria.

**1970s** — DDT is banned in most countries for its effects on the environment and health.

**2004** — The Stockholm Convention outlaws several products of the chemical industry, dubbed "persistent organic pollutants", which build up in the environment and can harm wildlife.

127

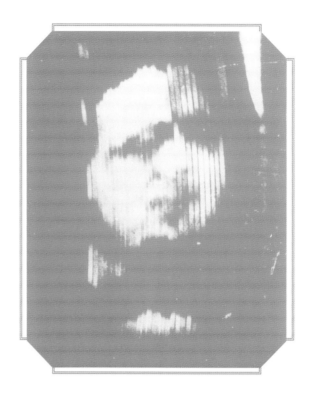

# Television

## TELEVISION CHANGED THE WAY OF LIFE OF HUNDREDS OF MILLIONS OF PEOPLE IN THE TWENTIETH CENTURY.

The history of this far-reaching invention is, however, far from simple. Dozens of inventive people contributed to its development. One of television's most significant pioneers was Russian-born inventor Vladimir Zworykin (1889–1982), who also made important contributions to the development of the electron microscope.

Zworykin was born in the town of Murom, in what was then the Russian Empire. As a child he spent time installing and repairing electric doorbells in the family-owned passenger steamships. In 1912, he obtained a degree in engineering from the Saint Petersburg Institute of Technology. At the Institute, one of Zworykin's professors, Boris Rosing (1869–1933), showed him a project he had been working on in secret. Rosing called it "electric telescopy" – one of the early names for television. Unbeknownst to either Zworykin or Rosing, however, several other people in other countries were working on the same idea.

Indeed, as early as 1908 the Scottish engineer AA Campbell

Swinton (1863–1930) had published a letter in which he outlined his concept for "distant electric vision" using the cathode-ray tube. Invented in 1897 by German physicist Karl Ferdinand Braun (1850–1918), the cathode-ray tube is a glass tube, from which the air has been removed. A beam of electrons strikes a flat screen, the inside of which is coated with chemical compounds called phosphors, which glow wherever electrons collide with them. Electromagnets positioned around the tube control the direction of the beam, and the signal fed

**ABOVE:** An image of John Logie Baird's business partner, Oliver Hutchinson (1891–1944), was transmitted to members of the Royal Institution in early 1926.

## TELEVISION PIONEERS

For much of the 1930s, Vladimir Zworykin was embroiled in a lengthy patent battle between the Radio Corporation of America (RCA) and another television pioneer, American inventor Philo T Farnsworth (1906–1971). Farnsworth won the battle – at great cost to RCA.

Another important figure in developing electronic television was Hungarian inventor Kálmán Tihanyi (1897–1947), whose work was crucial in making Zworykin's Iconoscope camera work. There was another approach to television besides the all-electronic system: the "electromechanical system". In 1924, Scottish inventor John Logie Baird (1888–1946; shown above standing by the railings) transmitted the first-ever television pictures. The earliest photograph of a television picture (opposite) shows Baird's business partner.

**LEFT:** Combined electronic television set and radio receiver, 1938, made by British company Pye. The 23-centimetre (9-inch) cathode-ray tube (CRT) screen is a descendant of Zworykin's kinescope, the first practical television display.

**OPPOSITE TOP:** Zworykin standing next to an early scanning electron microscope, around 1945. Zworykin did not invent the electron microscope, but he led a team that made important improvements in the device, which has revolutionized biology, medicine and materials science.

**OPPOSITE BOTTOM:** Zworykin's night-vision device, the snooperscope, photographed in 1944. The snooperscope was sensitive to infrared radiation – or "heat rays" – which warm-blooded animals (including humans) emit with greater intensity than non-living things, by virtue of their warm bodies. Zworykin's device helped soldiers in night-time conflicts during the Second World War.

After submitting an improved patent application in 1925, Zworykin demonstrated his television system. The images were dim and stationary, and his employers were unimpressed. He received a more favourable response when he showed it to the Radio Corporation of America (RCA) in 1929. Zworykin's camera, later dubbed the Iconoscope, would become the standard way of producing television pictures. Zworykin developed the technology further at the RCA. In 1939, the company demonstrated it at the New York World's Fair and, in 1941, the RCA began regular commercial television broadcasts in the USA.

Across the Atlantic, on the West Coast of Scotland, John Logie Baird had been watching the various early efforts to transmit moving pictures. Thanks to ill-health, something that would dog his entire life, Baird was rejected by the armed forces during World War One. He worked instead as superintendent engineer of the Clyde Valley Electrical Power Company.

After the war, he began working in earnest on his goal; television. Working, in true British style, with any bits and bobs he could find he finally managed, in 1924, to transmit a shaky image a few feet across a room. Instead of electron beams scanning the inside of a cathode ray tube, Baird's device used spinning discs with spiral holes to produce images. By 1926, he'd improved his apparatus enough to demonstrate it to 50 fellow scientists in London and, in 1927, his television managed to work across 438 miles of telephone line between London and Glasgow.

The Baird television Development Company transmitted the first transatlantic TV transmission in 1928. He even managed stereoscopic and colour images. Baird's device, however, was electromechanical, and, separately, Marconi-EMI were now developing electronic systems, slowly rendering Baird's work yet another casualty in the race to our living rooms.

to the magnets causes the beam to scan in horizontal lines across the screen. By scanning the whole screen in this way several times every second, while also varying the intensity of the electron beam, it is possible to display a moving image. Swinton never attempted to build the system he conceived, and Rosing's system was crude and unwieldy, but both men would prove important in the early hive of minds contributing to the birth of television.

In 1919, after the Bolshevik Revolution during the Russian Civil War, Zworykin emigrated to the USA. Within a year he had begun working at the Westinghouse Electric and Manufacturing Company in Pittsburgh. In 1923, after spending a considerable amount of his spare time working on television, he applied for a patent. Zworykin's system used one cathode-ray tube to display pictures and another one in the camera. Inside his television camera, light fell on the screen of the cathode-ray tube. Instead of phosphors, this screen was coated with light-sensitive dots made of potassium hydride. An electron beam scanned the screen, as in the picture tube, and each light-sensitive dot produced a signal that depended on the brightness of the image at that point.

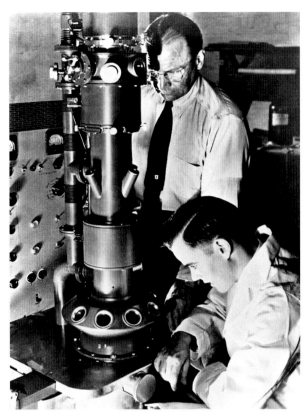

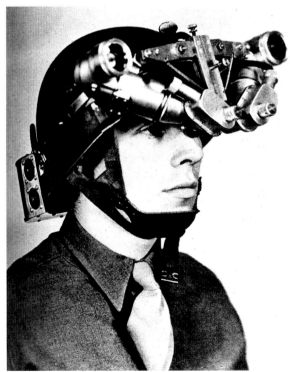

## TIMELINE

**1883** — German inventor Paul Nipkow (1860–1940) conceives an electromechanical system for transmitting live pictures: a rotating disc with holes in, called the Nipkow disk.

**1897** — German physicist Karl Ferdinand Braun invents the cathode-ray tube, which will become the mainstay of the television industry until the early 2000s.

**1909** — Russian engineer Boris Rosing is the first to use a cathode-ray tube as a television receiver.

**1923** — Zworykin applies for a patent for a "television system" which includes cathode-ray tubes for both the camera and the receiver (screen).

**1924** — Scottish inventor John Logie Baird becomes the first to transmit moving television pictures.

**1924** — Hungarian physicist Kálmán Tihanyi patents the "Rasioskop" image storage tube, RCA purchase the patent which enables them to dramatically increase the sensitivity of Zworykin's iconoscpe tube.

**1928** — American inventor Philo T Farnsworth is the first to demonstrate a working, all-electronic television system.

**1928** — Baird becomes the first to demonstrate colour television. Regular colour television broadcasts will not begin for more than 20 years.

**1929** — Regular television broadcasts begin in Germany and the UK. They are very low-definition broadcasts using mechanical scanning.

**1962** — The first television pictures are relayed by satellite, via the US satellite Telstar.

**2000s** — Plasma and LCD (liquid crystal display) televisions begin to displace cathode-ray tubes, accounting for more than 50 per cent of all TV sets by 2004.

**2000s** — Regular HD (High Definition) television broadcasts begin and viewers start to replace older "standard" television with the new technology.

**2010s** — 4K UHD TV becomes widely accepted.

# The Helicopter

A STRANGE AIRCRAFT TOOK TO THE AIR IN 1923.
IT WAS THE AUTOGYRO, AN AEROPLANE WITH BOTH A
PROPELLER AND A ROTOR.

Today, the autogyro is only flown by enthusiasts, having been superseded by the more manoeuvrable helicopter. The most important feature of helicopter design, however, the complicated mechanics at the hub of the rotor, was established in Spanish engineer Juan de la Cierva's (1895–1936) autogyros.

Juan de la Cierva was born to a wealthy family in Mercia, Spain. As a boy, he was inspired by the early pioneers of flight, and he became determined to be an aviator himself. In 1911, he went to study civil engineering in Madrid. That year, he and two friends experimented with gliders, and formed an aviation company. In 1912, Cierva built the first aeroplane in Spain, but during the following few years two of his aeroplanes crashed after stalling at low speed. As a result, he became determined to build an aeroplane that could not stall. He came up with the autogyro: an aeroplane with a propeller at the front and rotating wings – rotor blades – at the top. The rotor blades would always be moving fast relative to the air, and producing lift, even when the autogyro was moving slowly.

---

**ABOVE:** Juan de la Cierva (in front, piloting) in his C8 autogyro in September 1928, just before leaving Croydon Airfield, UK, en route to Paris, France. In Cierva's company's name, "Autogiro" was spelt with an "i", while the generic name for this kind of rotary-wing craft was "autogyro", with a "y".

133

Other inventors had experimented with rotors as early as 1907, but with little success. Cierva decided to leave his rotors unpowered, so that they would windmill or "autorotate" as the autogyro moved through the air. This approach had an added benefit: if the engine cut out, the autogyro would not crash to the ground. Instead, it would fall slowly, like a spinning sycamore seed case. In 1920, Cierva patented his idea, and tested small models of his autogyro concept. The models worked well, but when he scaled up his design, he found it had a tendency to flip over.

He soon realized why. As it turns, each rotor blade spends half the time moving forwards – into the oncoming air – and half the time moving backwards. This means that the advancing blade is moving through the air faster than the receding blade and so the lift force is greater on one side than the other.

Cierva looked back at his earlier models, and realized that the smaller rotor blades were flexible. As those rotors turned, the blades twisted slightly, automatically adjusting to the changing airspeed during each rotation, and producing constant lift. Cierva set about mimicking this phenomenon in his larger, metal blades. To do this, he incorporated a "flapping hinge" where each rotor blade met the rotor hub. In January 1923, Cierva's first successful prototype, the C4, flew 180 metres (200 yards) at an airfield near Madrid. This was the first stable flight of a rotating-wing aircraft in history, and was quickly followed by many longer, more sustained flights. In 1925, Cierva demonstrated autogyro C6 in England and, with the support of an investor, formed the Cierva Autogiro

**ABOVE:** A Pitcairn-Cierva autogyro taking off from the South Grounds of the White House in Washington, DC, in 1930. The aircraft has fixed wings, like an aeroplane, but most of the lift force is provided by the rotor blades. In 1933, Cierva dispensed with the fixed wings altogether.

**RIGHT:** Russian–American helicopter pioneer Igor Sigorsky, flying his VS-300 helicopter in 1940. The VS-300 was the first helicopter to have a tail rotor; until then, helicopters had two counter-rotating main rotors to keep them stable in flight. Both designs are still common today.

## HELICOPTERS

The autogyro, invented by Juan de la Cierva and later developed by Russian engineer Igor Bensen (1917–2000), was effective, safe, and moved through the air almost as fast as some aeroplanes. Autogyros found several uses during the Second World War, including reconnaissance and even the bombing of submarines. But autogyros could not hover, or perform truly vertical landings and take-offs so eventually helicopters gained the edge once they became practical.

It was Russian-American aviation pioneer Igor Sikorsky (1889–1972) who established the blueprint for the modern helicopter. Sikorsky built his first helicopter in 1909 but, as with other inventors' attempts at the time, it did not work. After working on fixed-wing aircraft during the 1910s and 20s, Sikorsky eventually produced one of the world's first successful helicopters, the VS-300, in 1939. He went on to design the first mass-produced helicopter, the Sikorsky R-4, in 1942. The overall layout of most helicopters has changed little since then.

Company. Three years later, Cierva flew his C8 autogyro from England to France. The C8 featured a "fully articulated rotor", with blades that could flex backwards to absorb the drag force (air resistance), which had previously caused some blades to snap.

More improvements followed, including a system to drive the rotor, just at take-off, so that the autogyro could rise vertically. The most obvious change came in 1933 when Cierva built autogyros with no wings and no tail. Up to this point, autogyros were controlled in the same way as fixed-wing aircraft: using moveable flaps on the wings and tail. This meant that pilots all but lost control at low speeds, so Cierva decided to find a way to control his autogyros by tilting the rotor. To do this, he had to design a complicated system of hinges and control levers around his rotor hub, and what he achieved formed the basis of all future helicopter rotors. Ironically, after devoting his career to avoiding the problems of stalling, Cierva was killed at Croydon airport, a passenger aboard a conventional fixed-wing aeroplane that stalled and crashed into a building just after take-off.

**OPPOSITE:** A Focke-Wulf Fw 61, the first fully controllable helicopter, in 1937. German engineer Heinrich Focke (1890–1979) designed this after working on Cierva autogyros. The pilot is German aviator Hanna Reitsch (1912–1979), who set many records, including being the first woman to fly helicopters.

**ABOVE:** A modern, fully articulated rotor. Each blade is able to move independently of the others, and can tilt to increase or decrease the lift force. Cierva developed the fully articulated rotor so that he could control his autogyros without fixed wings.

## TIMELINE

**1907** French bicycle manufacturer Paul Cornu (1881–1944) becomes the first person to lift off the ground in free flight in a rotary craft, in Lisieux, Calvados, in France. His flight lasted 20 seconds, and he reached an altitude of 30 centimetres (12 inches).

**1909** Russian–American helicopter pioneer Igor Sikorsky builds his first helicopter, the S-1, in Kiev, Ukraine. It is the first helicopter to have control features, but is not powerful enough to lift off.

**1920** Juan de la Cierva patents his autogyro.

**1922** Cierva conceives "flapping hinges", the key to successful rotor design.

**1923** Cierva makes the first successful sustained flight in a rotorcraft, in his C4 autogyro.

**1935** The first practical helicopter, Gyroplane Laboratoire, designed by French engineer Louis Charles Breguet (1880–1955), makes its maiden flight.

**1936** German engineer Heinrich Focke designs the first fully controllable helicopter, the Fw 61.

**1939** Sikorsky designs the VS-300, the first helicopter with a tail rotor.

**1942** Sikorsky's R-4 becomes the first mass-produced helicopter. By the end of the Second World War, Sikorsky will produce more than 400 of them, for the US Army.

**1945** Nikolay Ilyich Kamov (1902–1973) designs an ultralight helicopter, a "flying motorcycle" powered by a motorcycle engine.

**1965** Design of the Lockheed AH-56 Cheyenne, the first helicopter specifically created to carry attack weapons.

# Rockets

GERMAN-AMERICAN ROCKET ENGINEER WERNHER VON BRAUN DESIGNED THE FIRST ROCKET-POWERED LONG-RANGE BALLISTIC MISSILES – BUT HIS REAL ACHIEVEMENT WAS IN SPACEFLIGHT.

His determination in following his boyhood dream of sending people to the Moon, together with his excellent technical and leadership skills, made him the ultimate spaceflight pioneer of the twentieth century.

Wernher von Braun (1912–1977) was born a baron, to an aristocratic family in the town of Wirsitz, in the then German Empire (now Wyrzysk in Poland). After the First World War, his family moved to Berlin, Germany. Young Wernher became interested in space when his mother, a serious amateur astronomer, gave him a telescope – and he was mesmerized by stories of journeys into outer space. Von Braun studied mechanical engineering at the Charlottenburg Institute of

Technology, in Berlin. While there, he joined the Verein für Raumschiffahrt (VfR) – the Society for Spaceship Travel – and became involved in building and firing early liquid-fuel rockets.

Von Braun joined the German army's Ordnance Division in

ABOVE: Launch of a rocket bearing the Soviet space satellite, Sputnik 1 in 1957. This was the first man-made object to make it into orbit, and challenged the American's to follow suit just one year later.

137

**LEFT:** An A-4 rocket on a test launch at Peenemünde, Germany, in 1943. The A-4 became the V-2 when used during the Second World War. Payload: 1 tonne; maximum altitude: 95 kilometres (50 miles); maximum speed: 5,800 kilometres per hour (3,600 miles per hour); range: 320 kilometres (199 miles).

**ABOVE:** Officials of the US Army Ballistic Missile Agency at Redstone Arsenal in Huntsville, Alabama. Von Braun is second from right; in the foreground is Romanian rocket pioneer Hermann Oberth. The USA's first satellite, *Explorer*, was launched by a Jupiter-C rocket designed at ABMA.

October 1932, developing and testing rockets at an artillery range in Kummersdorf, near Berlin. He became technical head of the "Aggregate" programme, whose main aim was to design rockets for use as long-range ballistic missiles. In 1935, von Braun's team moved to Peenemünde, on the Baltic Coast, where the programme continued until the end of the Second World War, in 1945. Each rocket in the proposed Aggregate series was bigger and more ambitious than the last. For example, the A9/10, had it ever been launched, would have been a 100-tonne, two-stage rocket aimed at

## EARLY SPACEFLIGHT PIONEERS

It was when he read *Die Rakete zu den Planetenräumen* (*The Rocket into Interplanetary Space*) that Wernher von Braun set about learning the mathematics, physics and engineering necessary to make space travel a reality. The book was written by German rocket pioneer Hermann Oberth (1894–1989), in 1923.

Oberth was one of three visionaries who independently worked out how multi-staged rockets could be used to lift into space. The other two were Russian mathematics teacher Konstantin Tsiolkovsky (1857–1935) and American physicist Robert Goddard (1882–1945). In 1926, Goddard became the first person to build and fly a liquid-fuel rocket, on his aunt's farm in Massachusetts. In his day, Goddard was ridiculed in the press. Nevertheless, Wernher von Braun, although himself an innovator, based much of his early work on Goddard's research.

New York, United States; the A12 would have been a true orbital launch vehicle, able to place satellites into orbit.

The only Aggregate rocket to see service was the A-4, better known as the V-2. Designed by von Braun's team, this was the world's first medium-range ballistic missile – and the first reliable liquid-fuel rocket. By the end of the war, more than 3,000 had been launched; these terrible weapons, built by prisoners-of-war, rained destruction upon England, Belgium and France from 1944 onwards. Von Braun's involvement in the weapon's development and his membership of the Nazi party remain controversial, but he was always preoccupied with his real goal of sending rockets into space. When the war ended, the US Army took von Braun and his team of workers to the United States. In 1950, von Braun settled in Huntsville, Alabama, where he headed the US Army rocket team. At that time, the Cold War was intensifying, and the United States was worried that the Soviet Union might dominate the new territory of space. Throughout the 1950s, von Braun became something of a celebrity, promoting the idea of space travel in books, magazines, on television and in films – inspiring the American people with his dreams of space stations and journeys to the Moon and Mars.

The Space Age officially began on 4 October 1957, when the Soviet Union launched the first satellite, Sputnik 1, into orbit. The news prompted the United States Government to form NASA (the National Aeronautics and Space Administration). In 1958, a Redstone rocket, designed by von Braun, put America's first satellite into orbit. Two years later, NASA opened its Marshall Spaceflight Center, in Huntsville, and von Braun became its director. The Soviet Union got the

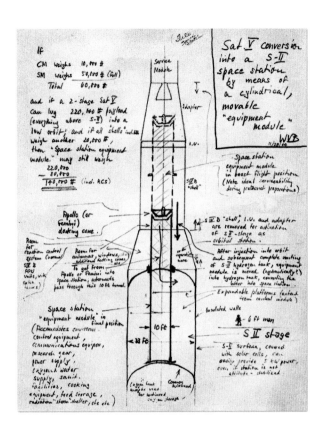

**800s**

Chinese alchemists develop "black powder", gunpowder, and use it to make "fire arrows", similar to modern firework rockets, for use in warfare.

**1200s**

The Monguls learn rocket technology after invading China; the Ottoman Turks later learn it from the Monguls and use it in Europe.

**1780s**

Tipu Sultan (1750–1799), Indian ruler of the Kingdom of Mysore, develops cardboard battle rockets with a range of up to 2 kilometres (1.2 miles).

**1804**

English inventor William Congreve (1772–1828) develops a rocket with a better range and accuracy than those used in Mysore.

**1903**

Russian teacher Konstantin Tsiolkovsky suggests that multi-stage liquid-fuel rockets could reach space, and works out the "rocket equation" for calculating a rocket's speed any time after launch.

**1923**

Romanian rocket pioneer Hermann Oberth's *Die Rakete zu den Planetenräumen*, its arguments similar to Tsiolkovsky's, inspires young Wernher von Braun.

**1926**

American inventor Robert Goddard builds and successfully launches the first liquid-fuel rockets.

**1932**

Wernher von Braun begins working for the German army's Ordnance Division, where he leads the development of rocket-powered ballistic missiles.

**1944**

A German V-2 is the first rocket to travel beyond space: to an altitude of 100 kilometres (60 miles), known as the Kármán Line after Hungarian-American physicist Theodore Kármán (1881–1963).

**1945**

American and Russian military personnel benefit from German wartime rocket expertise and develop their own missile and space programmes.

**1957**

A Russian R-7 Semyorka rocket launches the first man-made object into orbit: Sputnik 1, the first artificial satellite.

**1969**

An American Saturn V rocket launches the Apollo 11 astronauts into space en route to the first-ever Moon landing.

139

upper hand again in 1961, when it launched a human into space for the first time; the United States retaliated by launching Alan Shepherd into space less than a month later, again with a von Braun Redstone rocket.

In May 1961, to von Braun's delight, United States president John F Kennedy (1917–1963) announced the country's intention of "landing a man on the Moon and returning him back safely to the Earth". The United States succeeded – and the astronauts of the "Apollo" programme travelled to the Moon in modules launched into space atop huge Saturn V rockets, designed by von Braun's team at the Marshall Space Center. Von Braun had finally achieved his goal of interplanetary travel and NASA call him "without doubt, the greatest rocket engineer in history".

**OPPOSITE:** Russian space visionary Konstantin Tsiolkovsky, whose 1903 book *The Exploration of Cosmic Space by Means of Reaction Devices* was the first serious scientific treatise on using rockets to reach space.

**ABOVE:** Von Braun's hand-drawn design, 1964, explaining his design for how a space station could be carried in a Saturn V rocket. The USA's first space station was Skylab, which was launched in 1973 aboard a two-stage, rather than three-stage, version of the Saturn V.

**LEFT:** Wernher von Braun, in 1954, holding a model of a proposed rocket that would lift people into space. During the 1950s, von Braun was something of a celebrity in the USA, nurturing dreams of space travel among the postwar American people.

**OPPOSITE:** The launch of the Apollo 11 mission, 16 July 1969, carried into space by a huge Saturn V rocket from Cape Kennedy, USA. This was the realization of a childhood ambition for von Braun, who led the project to design and build the Saturn V.

## WOMEN ROCKETEERS

Though it was Wernher von Braun who won the plaudits as the brains behind US space rocketry, it was a lesser-known chemist who invented the fuel without which the Americans might never have made it into space at all. Mary Sherman Morgan (1921–2004) was America's first female rocketeer. The Hydyne-LOX propellant she created powered the Jupiter-C rocket that, on 31 January 1958, launched Explorer 1, the US's first space satellite.

Morgan's life began on a farm in rural North Dakota, where she lived until she ran away to college at the age of 19. Before finishing her degree, she was offered a job to work on explosives at the Plum Brook Ordinance Works in Ohio – then the biggest munition plant in the US. From there, she moved to the rocketry division of North American Aviation. She was the sole female scientist among the 900 rocketry experts working at the company, and one of only few without a college degree.

As a Theoretical Performance Analyst, Morgan's job was to calculate the expected performance of rocket fuels. Her task was to develop a new fuel powerful enough to launch a rocket into space, and stable enough for the rocket not to just explode on its launchpad. The fuel she invented was 60 per cent dimethylhydrazine and 40 per cent diethylenetriamine, mixed with LOX (liquid oxygen). It proved far more effective as a propellant than the fuel currently in use.

Morgan got little to no credit during her lifetime for her contribution to US rocketry, largely because the work she did was shrouded in Cold War secrecy, but she did not care about celebrity or being famous. It was only after her death that a one-time colleague spoke up for her, telling her son at her funeral that she had singlehandedly saved the American space programme from failure.

If Morgan played a vital part in getting the US into space, Alice King Chatham (1908–1989) made almost as significant a contribution. A sculptor by training, she was recruited by the US Air Force during the Second World War to devise a rubber oxygen mask for high altitude fighter pilots. She designed the pressure suits and helmets worn by the monkey-astronauts – the first living creatures the US sent into sub-orbital space – and went on to design the space helmets for the original seven Mercury astronauts. Chatham's other NASA inventions included a pressurized space suit and a space bed.

Fast forward the story to the 1990s, by which time NASA had sent astronauts to the Moon and was looking far beyond it to Mars and the further reaches of the Solar System. Aerospace engineer Donna Shirley (b.1941) was part of the team that designed heat shields for vehicles entering Mars's atmosphere, and then became head of the team that designed *Sojourner*, the first automated rover to explore the red planet's terrain and send pictures of its findings back to Earth.

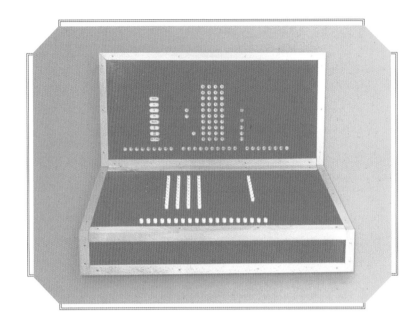

# Digital Computers

## THE FIRST ELECTRONIC DIGITAL COMPUTERS APPEARED IN THE 1940S. THEY WERE NOT SIMPLY THE RESULT OF ADVANCES IN ELECTRONICS.

Their development relied on a theory of computation formulated by English mathematician Alan Turing, who was also an important wartime code-breaker and a pioneer of machine intelligence.

Alan Turing was born in London to an upper-middle-class family, and his genius was evident from an early age. He taught himself to read in a matter of weeks and while in his teens at the auspicious Sherborne public school in Dorset he developed a fascination for science and mathematics. In 1931, he went to King's College, Cambridge, to study mathematics.

While he was at university, Turing became interested in logic. This was a hot topic in mathematics at the time: mathematicians were attempting to define their subject completely in terms of logic – to iron out inconsistencies and

to show that mathematics is "logically complete". In 1931, German mathematician Kurt Gödel (1906–1978) had published two theorems that showed this was impossible. He proved that

**ABOVE:** The "keyboard" of the Z3, a computer built in 1941 by German engineer Konrad Zuse (1910–1995). The Z3 was the first "stored-program" computer to use binary to represent numbers and instructions.

**OPPOSITE:** Pilot ACE, 1950. Towards the end of the Second World War, Turing told his colleagues he was "building a brain": the Automatic Computing Engine (ACE). After the war, Turing presented his design to the National Physical Laboratory. Pilot ACE was the prototype based on Turing's design.

even simple mathematical statements rely on assumptions and intuition that cannot be defined in terms of logic.

Inspired by Gödel's theorems, Turing wrote a landmark paper on the logic of mathematics in 1936. In this paper, Turing imagined an "automatic machine" that could read and write symbols on a tape, and carry out tasks based on a simple set of instructions. Turing proved that any problem that is "computable" can be solved by such a machine – a "universal" computer – if given the correct set of instructions. This was another way of expressing Gödel's theorems, since it also proved there were some mathematical statements that the machine could not compute. It was significant for another reason: Turing's hypothetical device became known as the "Universal Turing Machine" and was to be the blueprint for digital computers.

During the Second World War, Turing worked for the UK government helping to decode the German military forces' encrypted communications, at a Buckinghamshire mansion called Bletchley Park. The Germans used two devices, the Enigma machine and the Lorenz Cipher machine, to produce extremely well-encrypted communications. Although possible to find "keys" to crack the encryption, this was a laborious process. In the early 1930s, Polish code-breakers had built a machine that sped up the process. But in 1939, the Germans improved their machines, making the codes even harder to crack. Turing in turn designed a more efficient and faster machine, which he called "The Bombe". By the end of the war, 211 Bombes were operational, requiring 2,000 staff to run them. Turing's invention greatly helped the war effort, and probably shortened the war by a year or more.

After the war, he wrote a proposal to the National Physical Laboratory in London for an "automatic computing engine", based on his Universal Turing Machine. While his proposal was accepted, it was thought too ambitious, and a smaller version – the Pilot ACE – was built instead. It ran its first program in 1950. Other researchers were working on Turing Machines, too. The world's first stored-program, general-purpose computer was the Small Scale Experimental Machine, built by a team at the Victoria University of Manchester, also in England. It ran its first program in 1948.

Turing was well aware of the possibility that machines might one day "think". In an article in 1950, he suggested a

test for artificial intelligence: a person (the judge) would have two conversations via a keyboard and monitor – one with a human being and one with a computer. If the judge was not certain which was which, the computer would be deemed intelligent. No computer has yet passed the test.

In 1945, Turing was awarded the OBE (Order of the British Empire) for his work at Bletchley Park, but in 1952 he was convicted for homosexuality, then illegal in the UK (the UK government issued a posthumous apology to Turing in 2009). Two years after his conviction, he was found dead in his bed from cyanide poisoning; an inquest concluded that it was suicide.

**LEFT:** Alan Turing, photographed in 1951. No individual invented the computer, but Turing developed some fundamental theoretical and practical insights in the 1930s and 40s.

**BELOW:** A Colossus code-breaking computer at Bletchley Park, UK, 1943. Designed by English electronic engineer Tommy Flowers (1905–1998), the Colossus was the first fully electronic, stored-program computer – but it was not a truly general-purpose computer.

**OPPOSITE:** Grace Murray Hopper working on a manual tape punch computer, 1944.

144

## COMPUTER LANGUAGE AND MICROCHIPS

Among the many brilliant people who contributed to computing, two pioneers stand out. Grace Murray Hopper (1906–1992) was a groundbreaking computer scientist who devised the first code compiler to effectively convert English into a computable language. Her first compiler translated mathematical notation into machine code; the Common Business Orientated Language, or COBOL, she devised in 1958 is still in use around the world today. In the 1980s, Lynn Conway (b.1938) pioneered the microchip revolution that made modern personal computers and smartphones possible. Her rules standardized microchip design, making the process faster, easy to scale down and more reliable.

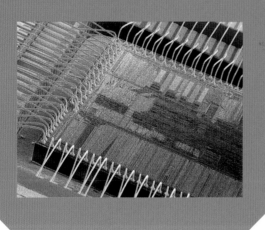

# TIMELINE

**1936** — Turing writes his landmark paper, "On Computable Numbers, with an Application to the Entscheidungsproblem", which has great influence on the development of computers.

**1937** — American electronic engineer Claude Shannon (1916–2001) designs circuits that carry out "logical" operations using binary numbers; his approach influences the design of electronic computers.

**1941** — The Z3, built by Konrad Zuse, is the world's first programmable automatic computing machine. It is an electromechanical machine, using relays as switches, rather than valves.

**1942** — American physicist John Atanasoff (1903–1995) and American engineer Clifford Berry (1918–1963) build the ABC: the world's first all-electronic, digital – but not general-purpose – computer.

**1943** — 1940s The first all-electronic, fully programmable, general-purpose computers appear, such as ENIAC, EDVAC and Pilot ACE.

**1945** — John von Neumann submits his "First Draft of a Report on the EDVAC", which contains an influential description of the "stored program" computer architecture.

**1950** — American electrical engineer J. Presper Eckert (1919–1995) develops the first "read-write" computer memory: the mercury delay line.

**1954** — Core memory, made of an array of tiny magnetic rings on criss-crossed wires, provides the first real "random access memory" (RAM).

**1954** — Transistors begin to replace valves, making electronic computers smaller, faster and much less power-hungry.

**1956** — The RAMAC 350, by American company IBM, is the first hard disk. It stored less than four megabytes of information, but weighed more than a tonne and would only just fit through a doorway.

**1971** — American company Intel releases the first microprocessor – a complete CPU on an integrated circuit, or "chip".

**RIGHT:** It was six mathematicians – Jean Jennings, Betty Snyder, Kathleen McNulty, Marlyn Wescoff, Francis Bilas and Ruth Lichterman – who devised the software needed to run ENIAC (Electronic Numerical Integrator and Computer), built for the US Army Ballistic Research Laboratory. They also compiled the manual on how to work the machine. The programmers started by breaking down differential equations into their constituent parts and then calculating how long it would take ENIAC to solve each piece. The relentless trial and error this involved meant they not only had to set dozens of dials, but plug multiple cables into the front of the machine. Despite the vital importance of their work, their contribution was ignored for more than 50 years after ENIAC ran for the first time in February 1946.

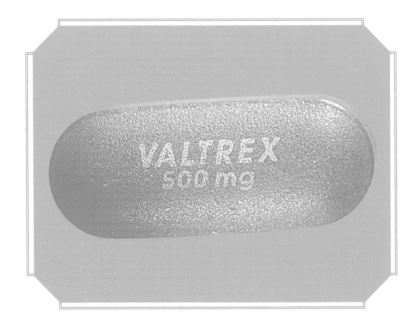

# Antiviral Drugs

THE INVENTIONS OF AMERICAN BIOCHEMIST
GERTRUDE ELION ARE FAR TOO SMALL TO SEE. THEY
ARE WORKS OF ENGINEERING, BUT AT THE MOLECULAR
LEVEL: ELION WAS A PIONEER OF CHEMOTHERAPY.

The medicines she developed have brought hope to millions of people with bacterial and viral infections and cancer.

Gertrude Elion (1918–1999) was born in New York, USA. Her mother was from Russia, her father from Lithuania. As a child, "Trudy" had an insatiable desire to read and learn, and she took an interest in all subjects. It was the fact that her grandfather had died of leukaemia that fostered her interest in science. At the age of 15, she began studying chemistry at Hunter College, New York, in the hope that she might one day develop medicines to cure or prevent the disease that had claimed her grandfather.

The campus at Hunter College was for women only, so Elion was used to women studying science. However, in the world outside college, men still dominated, and despite her outstanding academic record, Elion found it impossible to get funding to take on a PhD. By doing several poorly paid jobs, she managed to save up enough money to enrol at night school, and she received a masters degree in 1941, but never received a PhD. That year, many men were out of the country fighting in the Second World War, so some laboratories were employing women.

**ABOVE:** Today there are dozens of anti-viral drugs available, including this one, Valtrex®. The active ingredient in this drug is a derivative of acyclovir, developed by Elion. Valtrex® is used to treat all kinds of herpes infections, including genital herpes, shingles and AIDS-related herpes.

**BELOW:** Gertrude Elion and George Hitchings, photographed shortly after winning the Nobel Prize for Medicine, in 1988. In 1991 Elion became the first woman to be inducted into the US National Inventors Hall of Fame. Elion worked closely with Hitchings for much of her career.

**RIGHT:** Replica of DNA model originally assembled by English biologist Francis Crick (1916–2004) and American molecular biologist James Watson (b.1928) in 1953. Along the length of each helical strand are the purines and pyrimidines, the key to most of Gertrude Elion's inventions.

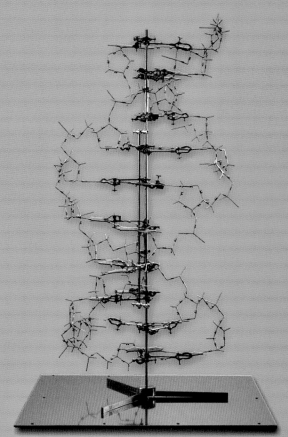

In 1944, after several years of working in unchallenging jobs in the chemical industry, Elion began work as a senior research chemist in the New York laboratory of the pharmaceuticals company Burroughs Wellcome. There she worked as an assistant to American doctor and chemist George Hitchings (1905–1998), who encouraged her to learn as much as possible and to follow her own lines of enquiry.

Although Elion had studied chemistry, her quest to produce medicines had led her to biochemistry (the chemistry of living things), pharmacology (the study of how drugs work) and virology (the study of viruses). By the 1940s, biochemists had discovered that a chemical called DNA (deoxyribonucleic acid) present in the cell nucleus was involved in cell replication. They had worked out the constituent parts of DNA, but its double helix structure would not be worked out until 1953. The most important constituents are small molecules called purines and pyrimidines, which join together in pairs along the length of the much larger DNA molecule. Elion wondered whether altering these molecules might somehow confuse a virus or a bacterium or stop the uncontrolled reproduction of cancer cells. So she and Hitchings set about engineering new ones.

Elion and Hitchings made their first breakthrough in 1948. One of their purines, 2,6-diaminopurine, was found to restrict the reproduction of bacteria, and to slow the growth of tumours in mice. Over the next few years, Elion tested more than 100 other engineered purines. In 1951, trials in rats suggested that one of them, 6-mercaptopurine (6-MP), could fight leukaemia. At the time, there was little hope for patients with leukaemia, most of whom were children and most of whom died within a few months of diagnosis. When 6-MP was tested in humans, it was found to increase life expectancy, and some children even went into full remission. The drug is still used today in anti-cancer chemotherapy.

With increasing knowledge of biochemical reactions at the heart of cell biology, Elion went on to synthesize several medicines effective against a range of bacterial diseases, including malaria, meningitis and septicaemia. In 1958, she produced the first medicine that could suppress the immune

## TRANSPLANT SURGERY

In 1958, American doctor William Dameshek (1900–1969) suggested that Gertrude Elion's anti-leukaemia drug 6-MP might be effective at suppressing the immune system. If true, the drug might prevent the body's rejection of organs after transplant surgery. Damashek's rationale was that the white blood cells responsible for the immune response were similar to the white blood cells involved in leukaemia.

In 1960, English transplant pioneer Roy Calne (b.1930) tested 6-MP in dog kidney transplants, and found that it was fairly effective. Gertrude Elion suggested that a related compound, azathioprine, might be more effective, and Calne conducted promising trials with the new drug in 1961. The first successful kidney transplant between unrelated humans was performed soon after, using azathioprine. In combination with corticosteroids, this drug became the mainstay of transplant surgery, until it was replaced by a more powerful drug, cyclosporine, in 1978.

system, making organ transplants safer (see box). In 1981, after more than a decade's work, she created the first anti-viral drug, acyclovir, which is the active substance in anti-herpes medicine such as Zovirax® and Valtrex®. Gertrude Elion received many awards for her groundbreaking work in chemotherapy, including, in 1988, the Nobel Prize in Physiology or Medicine. She shared the prize with George Hitchings and Scottish pharmacologist James Black (1924–2010), for "discoveries of important principles for drug treatment".

In 1968, Elion had hired organic chemist Janet Rideout (b.1939) at her pharmaceutical firm Burroughs Wellcome. Rideout's achievements include the development of AZT (azidothymidine), the first effective anti-AIDS drug. It was a breakthrough that, over the years, has saved millions of lives. She also helped to develop acyclovir, the first effective treatment for the herpes simplex virus.

Rideout started work on AZT – which had first been synthesized as a possible anti-leukaemia drug in 1964 – in June 1984. It was one of the many compounds she tested, but the only one that seemed to demonstrate potential. What Rideout discovered was that, though AZT did not work on common viruses, it was very effective against the enzymes that certain retroviruses, such as HIV, use to reproduce. By the end of 1984,

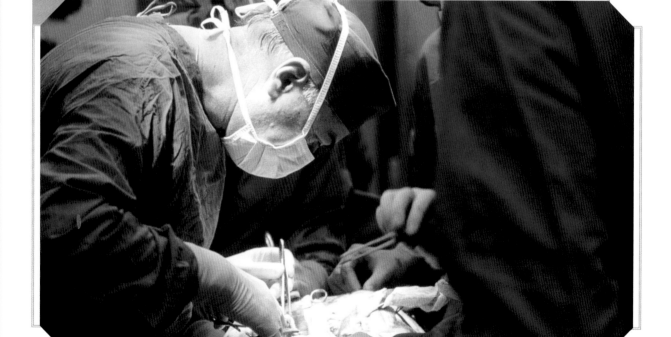

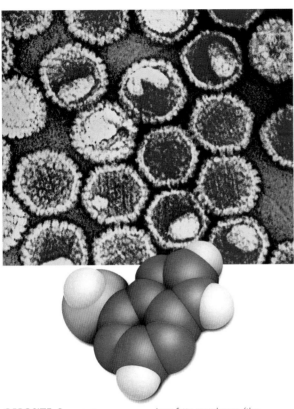

**OPPOSITE:** Surgeons removing and preparing a human kidney in preparation for transplantation into a recipient. Since the development of immunosuppressive drugs – such as azathioprine, developed by Elion – there is no need for the donor and recipient to be related to avoid rejection of the organ by the recipient's body.

**ABOVE,** top: False-colour electron micrograph of herpes simplex viruses. Each virus comprises DNA in a protein "cage" (the capsid), surrounded by a fatty membrane (the envelope). A virus uses resources inside a host cell to reproduce; Elion produced the first effective anti-viral drugs, which inhibit this process.

**ABOVE,** bottom: Molecular structure of 6-mercaptopurine (6-MP), developed by Elion and Hutchings in 1951. It has a very similar shape to purine molecules found along the length of DNA, and interrupts their formation, inhibiting the rampant reproduction of DNA characteristic of cancers.

she had demonstrated that AZT could be used to treat two animal retroviruses effectively.

The next step was to see if what worked with rats would work on human cells. Tests at the National Cancer Institute proved AZT's efficacy. In 1987, the Federal Drugs Agency approved AZT for human administration; the following year, it was patented in Rideout's name along with four of the other scientists who had collaborated with her.

|  |  |
|---|---|
| 1891 | Emil von Behring (1854–1917) develops a successful antitoxin against diphtheria. |
| 1899 | German chemical company Bayer begins selling chemically derived acetylsalicylic acid as aspirin. |
| 1905 | German microbiologist Paul Ehrlich develops the first purely synthetic pharmaceutical – what he calls a "magic bullet". The chemical, Salvarsan, is the first effective treatment against syphilis. |
| 1940s | Australian pharmacologist Howard Florey (1898–1968) heads a team that pioneers the clinical use and mass-production of penicillin, the first effective antibiotic. |
| 1950s | Mexican pharmaceutical company Syntex develops the first effective oral contraceptive pill, which contains progestin, a synthetic hormone that reduces the frequency of ovulation. |
| 1951 | Gertrude Elion and George Hitchings develop 6-mercaptopurine as one of the earliest chemotherapeutic agents. |
| 1953 | American pharmaceutical company Stirling-Winthrop begins selling paracetamol. |
| 1963 | Diazepam, a treatment against anxiety and insomnia, is marketed as Valium, and quickly becomes one of the most successful pharmaceuticals in history. |
| 1965 | Syrian-born biologist Roger Altounyan (1922–1987) develops the first effective antihistamine drug, sodium cromoglicate, as a treatment for asthma. |
| 1978 | American biochemist Herbert Boyer (b.1936) develops a way to engineer bacteria to produce human insulin, increasing the availability of this anti-diabetes drug. |
| 1981 | Elion produces acyclovir, the first effective anti-viral drug. |

151

# The World Wide Web

### IN MODERN LIFE, IT SEEMS INCREASINGLY HARD FOR AN INDIVIDUAL TO INVENT SOMETHING THAT TRULY CHANGES THE WORLD.

However, one person who did just that is English physicist and computer scientist Tim Berners-Lee (b.1955). In 1990, he launched the World Wide Web.

Timothy Berners-Lee was born in London. His parents were both computer scientists. As a boy, Tim became interested in electronics after building circuits to control his model train set. He studied physics at Oxford University; while he was there, he built his first computer. After graduating in 1976, he worked as a computer systems engineer at various companies.

In 1980, Berners-Lee spent six months at the European Organization for Nuclear Research, a particle physics facility in the outskirts of Geneva, on the border between France and Switzerland. It is better known by the acronym CERN, which derives from the facility's original name, Conseil Européen pour la Recherche Nucléaire. While at CERN, Berners-Lee

devised a computer system, for his own personal use, to store and retrieve information. Named ENQUIRE, this was a forerunner of the Web. It was based upon hyperlinks, cross-references in one document that enable a computer to call up another, related document.

In 1984, Berners-Lee was back at CERN, on a computing fellowship programme. He became frustrated by the lack of compatibility between different computer systems, and between documents written using different software applications. In a memo he sent to his manager in 1989, Berners-Lee set out his vision of a "universal linked information system" with which to organize the huge amounts of information produced at CERN. He proposed that a "web of links" would be more useful than the "fixed, hierarchical system" that existed. Documents available on computers

**LEFT:** During the 1960s, most large businesses and universities had a centralized "mainframe" computer like this. Computer networking, upon which the Web depends, originated in efforts to establish time-shared access to these machines via terminals distributed through the organization.

**BELOW:** The actual "NeXTcube" computer Berners-Lee used to host the first web page, and to write the software necessary to implement his idea. The computer was connected to the local network at CERN. A sticker on the processing unit reads: "This machine is a server: DO NOT POWER DOWN!"

within CERN's network would contain hyperlinks to other documents, including those on different computers. In 1990, Berners-Lee's manager encouraged him to spend some time – as a side project – on developing his idea.

During the autumn of 1990, Berners-Lee, along with his colleague, Belgian computer scientist Robert Cailliau (b.1947), created all the now-familiar fundamental components of the World Wide Web. The universal language he invented for writing linked documents (web pages) is "html" – hypertext markup language. The software that responds to "requests" from hyperlinks is called a "web server", a term that also refers to the hardware that hosts the web pages. And the language, or protocol, computers use to communicate the hyperlink requests is "http" – hypertext transfer protocol. Berners-Lee had to write the first web browser, the application used to view the documents hosted on web servers. He called his browser "WorldWideWeb". Berners-Lee also wrote the first web pages, which he published on his server in December 1990. It was on 25th of that month that Berners-Lee first "surfed" from one web page to another, via http, by clicking a hyperlink in his browser.

The following year, Berners-Lee made available his software to people outside CERN, and the idea quickly caught on. By 1994, the Web had grown so much that each "resource" – a document or image, for example – needed a unique "address" on the Internet. In consultation with the Web

## WHERE THE
# WEB
## WAS BORN

In the offices of this corridor, all the fundamental technologies of the World Wide Web were developed.

Started in 1990 from a proposal made by Tim Berners-Lee in 1989, the effort was first divided between an office in building 31 of the Computing and Networking Division (CN) and one in building 2 of the Electronics and Computing for Physics Division (ECP).

In 1991 the team came together in these offices, then belonging to ECP. It was composed of two CERN staff members, Tim Berners-Lee (GB) and Robert Cailliau (BE), aided by a number of Fellows, Technical Students, a Coopérant and Summer Students.

At the end of 1994 Tim Berners-Lee left CERN to direct the WWW Consortium (W3C), a world-wide organization devoted to leading the Web to its full potential. The W3C was founded with the help of CERN, the European Commission, the Massachusetts Institute of Technology (MIT), the Institut National pour la Recherche en Informatique et en Automatique (INRIA), and the Advanced Research Projects Agency (ARPA).

In 1995 Tim Berners-Lee and Robert Cailliau received the ACM Software System Award for the World Wide Web. In 2004, Tim Berners-Lee was awarded the first Millenium Technology Prize by the Finnish Technology Award Foundation.

*The CERN Library*
*June 2004*

community, Berners-Lee created the format for web addresses, called the "uniform resource locator" (URL). After 1994, the Web spread rapidly beyond academic and military circles. Within a few short years, most people in the world had been affected directly by its existence, and millions were already regularly "surfing" from document to document online.

Tim Berners-Lee has received a huge number of accolades for his invention, which he gave free to the world without patents or rights. In 1994, he founded the World Wide Web Consortium, which helps keep the Web working smoothly and aims to foster its future growth. He also campaigns to keep the Internet "neutral" – free of restrictions on content and what kinds of computers may be connected.

**ABOVE:** A plaque at CERN commemorating the invention of the Web.

**RIGHT:** Tim Berners-Lee, who invented the World Wide Web in 1989, while working at CERN, in Geneva. When he left CERN in 1994, he created the World Wide Web Consortium (W3C), an organization that sets and maintains standards for the Web.

## DOUG ENGELBART
## (1925–2013)

Two very important technologies underpinned Tim Berners-Lee's invention of the World Wide Web: hyperlinks and the computer mouse. American computer scientist Douglas Engelbart invented the mouse in 1967, and he was also heavily involved in the development of hyperlinks.

In the 1960s, Engelbart headed a team at the Augmentation Research Center, at the Stanford Research Institute, California. Engelbart's team devised an online "collaboration system" called NLS (oN-Line System). This included the first use of hyperlinks and the mouse, which Engelbart invented in 1967. In 1968, Engelbart demonstrated NLS to a large audience of computer scientists. In addition to hyperlinks and the mouse, the 90-minute session, normally referred to as "The Mother of All Demos", introduced such ideas as email, video-conferencing and real-time collaboration between computer users far apart.

## TIMELINE

**1962** Doug Engelbart develops NLS, the first system to use hyperlinks.

**1962** Polish-American computer scientist Paul Baran (1929–2011) invents packet switching: breaking computer files into "packets" for effective transmission across a network.

**1969** The US government's Advanced Research Projects Agency connects several academic institutions together to form ARPANET, the beginning of the Internet, in which the Web exists.

**1971** Computer scientists at the American research centre Xerox PARC develop the first "graphical user interface" which implements WIMP (windows, icons, menus and a pointer).

**1972** American programmer Ray Tomlinson (1941–2016) writes the first e-mail program to send messages across ARPANET, and decides to use the "@" symbol to delineate the user's name from the server's name.

**1989** American company The World, based in Boston, becomes the first public Internet service provider, offering anyone dial-up access to the Internet.

**1990** Berners-Lee invents the World Wide Web.

**1993** Dutch-Canadian computer scientist Oscar Marius Nierstrasz (b.1957) produces the first web search engine.

**1994** Berners-Lee founds the World Wide Web Consortium.

**1998** College students Larry Page (b.1973) and Sergey Brin (b.1973) create the "PageRank" algorithm, the basis of US company Google.

**1998** The Internet boom begins: thousands of commercial Internet companies, known as dotcoms, attract unrealistic investment. Three years later, the "dotcom bubble" bursts.

**2000s** A new wave of developments – including blogging, video sharing, wikis and social networking – brings greater collaboration, referred to as "Web 2.0".

155

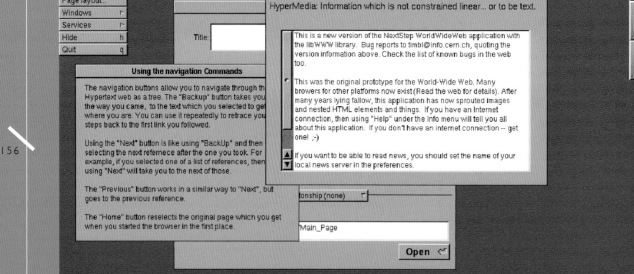

**156**

**ABOVE:** A 1994 screenshot of the first web browser, World Wide Web. Berners-Lee wrote the software exclusively for NEXT computers, like the one he used at CERN. The software could read and edit pages written in html, open linked pages and download any linked computer files.

**OPPOSITE:** The first page of the historic proposal for "Information Management" at CERN, submitted by Berners-Lee in March 1989, to his boss Mike Sendall (1939–1999). The words "vague but exciting" were written (top right) by Sendall, who encouraged Berners-Lee to spend some time on his idea the following year.

*Vague but exciting ...*

CERN DD/OC             Tim Berners-Lee, CERN/DD

**Information Management: A Proposal**       March 1989

# Information Management: A Proposal

## Abstract

This proposal concerns the management of general information about accelerators and experiments at CERN. It discusses the problems of loss of information about complex evolving systems and derives a solution based on a distributed hypertext sytstem.

Keywords: Hypertext, Computer conferencing, Document retrieval, Information management, Project control

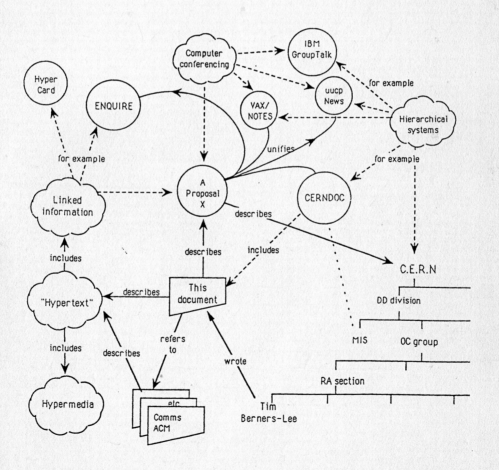

# Index

(Page numbers in **bold** refer to main entries, including accompanying photographs and illustrations, *italic* to other illustrations, photographs and captions, **bold italic** to timelines)

# Acknowledgements

The publisher would like the thank the following for their help with the production of this book:

**Science Museum:** Deborah Bloxham, Tom Vine, Ben Russell, Andrew Nahum, John Liffen, Alison Boyle, David Rooney, Jane Insley, Tilly Blyth, Yasmin Khan, Jasmin Spavieri, Doug Millard, Selina Hurley
**Science and Media Museum:** Colin Harding, Michael Harvey, John Trenouth, Philippa Wright
James Wills

# Credits

The publishers would like to thank the following sources for their kind permission to reproduce the pictures in this book.

Key: t = top, b = bottom, c = centre

Alamy: 9t, 16, 17; /Theo Alers 119; /Bygone Collection 118; /Gavin Hellier 14t; /Granger Historical Picture Archive 41; /Hemis 24; /Interfoto 142; /LondonPhotos/Homer Sykes 33b; /North Wind Picture Archives 21b; /Science History Images 94, 145b

Bridgeman Images: 89t

British Library: 21t

CERN: 154t, 156, 157

Luigi Chiesa via Wikimedia Commons: 38l

Rik Ergenbright: 68

Getty Images: 9b, 144t, 153t; /Archiv Gerstenberg/ullstein bild 8, 32; /Apic 60, 155; /Bettmann 20, 29t, 40, 48, 50t, 74, 83, 85, 96, 97t, 101b, 103t, 103c, 120, 122t, 129b, 134t; /Corbis 131t; /Fox Photos 91; /E. O. Hoppe/Mansell/The LIFE Picture Collection 126t; /Hulton Archive 107t, 114; /Hulton-Deutsch Collection/Corbis 51c, 67r, 132; /Karjean Levine 154b; /Will And Deni Mcintyre/The LIFE Images Collection 149l; /Mondadori Portfolio 111b; /George Rinhart/Corbis 133; /Vittoriano Rastelli/Corbis 150; /Schenectady Museum; Hall of Electrical History Foundation/CORBIS 84b; /Sotheby's/BWP Media 49; /SSPL 95; /Three Lions 125r; /Time Life Pictures/Mansell/The LIFE Picture Collection 72

Library of Congress, Washington: 112

Mary Evans Picture Library: 34b, 113t

David Monniaux: 135b

NASA: 31b, 127, 140-141

Nikola Tesla Museum, Belgrade: 104

© Parks Canada/Alexander Graham Bell National Historic Site of Canada: 92-93

Private Collection: 12, 23, 33t, 34t, 64b, 77t, 105, 114l, 125l

Public Domain: 101t

Science Photo Library: 19b, 121t; /AMI Images 151t; /Dr Tim Evans 151c; /Power & Syred 145t; /Gianni Tortoli 31t; /Science Source 19t; /Leonard Lessin 148; /Los Alamos National Laboratory 146-147; /Peter Menzel 102; /Sputnik 138b; /Shelia Terry 10, 25t, 44t, 121b; /John Walsh 98t

Science and Society Picture Library: 13b, 14b, 15, 23t, 23b, 24, 26, 28, 29b, 30, 35, 37c, 39, 42, 43, 44bl, 45, 46, 47, 50b, 51b, 52-55, 56, 57, 58, 59, 61, 62, 63, 65, 66, 67l, 68t, 69, 73t, 73b, 77b, 80, 81, 82, 86, 89b, 90, 97b, 98b, 99, 106, 107b, 108, 109, 110, 111t, 113b, 114, 116-117, 122b, 128, 129t, 130, 137l, 138t, 139, 143, 144b, 149r, 153b

Shutterstock: Everett Historical 37b; /Fotokostic 126c; /Gianni Dagli Orti 25b; /Nexus7 38r; /Priceless-Photos 152; /Somchai Som 87; /Sonny Meddle 98c; /Sovfoto/Universal Images Group 136; /Rudmer Zwerver 84t

Topfoto: 79, 88, 131b; /Granger Collection 13t, 23tr, 75, 76, 123, 137r; /ullstein bild 18, 124, 126b, 134b

Wikimedia Commons: 10b

Every effort has been made to acknowledge correctly and contact the source and/or copyright holder of each picture and Carlton Publishing Group apologises for any unintentional errors or omissions, which will be corrected in future editions of this book.

# Further Information

**Science Museum,** www.sciencemuseum.org.uk
**Science and Media Museum,** www.scienceandmediamuseum.org.uk
**Library of Congress,** www.loc.gov
**National Archives of the United States of America,** www.archives.gov
**Nikola Tesla Museum, Belgrade,** www.tesla-museum.org
**The Royal Institution,** www.rigb.org

# Publishing Credits

Editorial Director: Piers Murray Hill
Editor: Georgia Goodall
Additional Editorial: Gemma Maclagan, Lesley Malkin, Jane McIntosh, Catherine Rubinstein, Sandra Lawrence and Jeremy Harwood
Design Direction: Natasha Le Coultre
Memorabilia: Andrew McGovern
Picture Research: Steve Behan